Augustus D. Waller

Lectures on Physiology

On animal electricity

Augustus D. Waller

Lectures on Physiology
On animal electricity

ISBN/EAN: 9783337237851

Printed in Europe, USA, Canada, Australia, Japan

Cover: Foto ©berggeist007 / pixelio.de

More available books at **www.hansebooks.com**

LECTURES ON PHYSIOLOGY

FIRST SERIES

ON ANIMAL ELECTRICITY

BY

AUGUSTUS D. WALLER, M.D., F.R.S.

Fullerian Professor of Physiology at the Royal Institution of Great Britain
Lecturer on Physiology at St. Mary's Hospital Medical School,
London

LONGMANS, GREEN, AND CO.
39 PATERNOSTER ROW, LONDON
NEW YORK AND BOMBAY
1897

JOHN BALE SONS and DANIELSSON
GT. TICHFIELD STREET
LONDON

PREFACE.

The following "First Series" of lectures, as now printed, contain part of the material of a course of twelve lectures on "Animal Electricity," delivered at the Royal Institution during the spring of 1897.

The six printed lectures are by no means co-extensive with the twelve spoken lectures. The former, more especially the fifth and sixth, include matter that I should have felt it impossible to consider at length in the face of a non-technical audience, but that I nevertheless regard as essential to the further study of the subject. The latter were of necessity largely diluted with elementary explanations, and included three lectures on the electromotive action of the heart, and on the action of nitrous oxide, the publication of which is reserved for a "second series."

In Physical Science the Royal Institution of Great Britain has played a useful as well as a brilliant part as an organ of public instruction; and the great physicists, who fashioned the future destinies of the Institution, did not neglect to cultivate that portion of physical science which is the domain of the physiologist—the physics of living matter.

Davy, by his close physiological study of nitrous oxide, pointed out the way towards the establishment of Anæsthesia. Faraday took part in the foundation of our knowledge of Animal Electricity by his physiological investigation of the electric eel. It is indeed through its Fullerian Professors of Chemistry rather than through its Fullerian Professors of Physiology, that the Royal Institution has furthered our knowledge of living matter.

No clearer proof could well be offered of the absolute dependence of any branch of Science upon its laboratories rather than upon its lecture-theatre. The Fullerian professorship of Chemistry has been fruitful, not only in its own department, but also in the allied department of Physiology. The Fullerian professorship of Physiology has been comparatively sterile, even within its own nominal province. Both chairs have been held by men of the highest distinction; but the former has rested upon a laboratory, while the latter—so far from resting upon a laboratory—does not possess even one small room in which to keep itself alive.

A. D. WALLER.

Weston Lodge, 16, Grove End Road, N.W.
August, 1897.

CONTENTS.

LECTURE I.

LECTURE II.

LECTURE III.

LECTURE IV.

LECTURE V.

ELECTROTONUS.

LECTURE VI.

ELECTROTONUS *(continued).*

ILLUSTRATIONS.

LECTURE I.

ANIMAL ELECTRICITY.

" Wonderful as are the laws and phenomena of
electricity when made evident to us in inorganic
or dead matter, their interest can bear scarcely
any comparison with that which attaches to the
same force when connected with the nervous system and with life."—FARADAY. "*Experimental
Researches in Electricity.*" *Fifteenth series,* 1844.

LECTURE I.

Currents of animal electricity are produced by animal voltaic
  couples, in which injured or active protoplasm is electro-
  positive ("zincative"), resting protoplasm electro-negative
  ("zincable").

The utilisation of a nerve as a test-object representative of living
  matter.

Experiments.—Current of a voltaic couple. Current of injury of
  muscle—its negative variation. Current of injury of nerve
  —its negative variation. Parallelism between mechanical
  and electrical effects. Electrical effects of non-electrical
  stimuli. Action of ether and of chloroform upon isolated
  nerve.

THE master-key to many otherwise most intricate
and complex problems of animal electricity is a very
simple idea. Active matter is electropositive[1] to in-

---

[1] This nomenclature, which is the opposite to the conventional denotation, will perhaps be found justified by the context

I

active matter; more active matter is electropositive
to less active matter; matter that is by any means
stirred up to greater activity is rendered electro-
positive towards undisturbed matter, matter whose
action is lowered is electronegative to matter whose
action is normal.

Picture to yourself a uniform mass or strand of
protoplasm, that is to say of living matter, inactive at

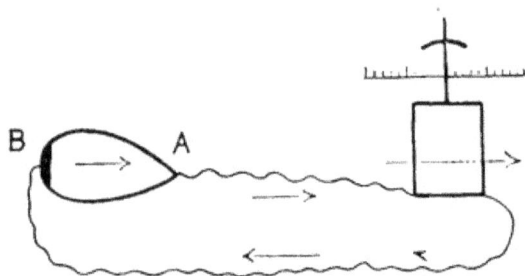

FIG. 1.—A lump of protoplasm as a voltaic couple.
Any injured, *i.e.*, chemically active spot B is zincative to any uninjured spot
A.   Current in the lump is from B to A, in the galvanometer from A to B.

of these lectures. In the phraseology generally employed by
physiologists the active spot is said to be " negative," and the
term "negativity of action " is derived from this.   But more
correctly speaking the active spot is positive, and we should
properly say " positivity of action."   But to reverse these and
other derived terms in common use would lead to hopeless confusion, which I desire to escape by employing the terms zincative,
zincativity.   Moreover we shall soon experience the want of a
word to denote that a resting spot, capable of activity, is capable
of becoming positive, or has a capability for becoming positive ;
this will be met by the words zincable, zincability, which are by
no means to be taken as synonymous with the terms excitable,

all points, or what comes to the same thing, equally active at all points. Any two points being *ex hypothesi* equally active, are equally electromotive—"isoelectric,"—and if connected by wires to a galvanometer, exhibit no current. But stir up B by pinching or pricking or by a touch of a hot wire, and you will at once obtain a current through the galvanometer that indicates the presence of current in the rest of the circuit as shown by the arrows. In the mass of protoplasm which is no longer uniformly active throughout, but more active at B than at A, there is current from B to A. In the galvanometer the current is from A to B.

These two unequally active regions B and A form a weak voltaic couple, of which B, the more active spot, where more chemical action is going on (we shall further on inquire into the possible character of such chemical action), is the generating or electro-

excitability. A spot of tissue under the influence of the anode is less excitable and more zincable. I have not been willing to use the less inelegant words electromotive and electromobile in place of zincative, zincable, for fear of an ambiguity in the term electromobile which has seriously impaired the precision of the term excitable. A tissue may be more excitable (erregbar) inasmuch as it may be aroused to action by a weaker stimulus, or more excitable (leistungsfähig) inasmuch as greater action is aroused by a given stimulus. A more zincable spot, as the name suggests, is capable of greater electropositive action than a less zincable spot.

positive plate, and A, the less active spot, where less
chemical action is going on, the collecting or electronegative plate.

Let me try to fix this point in your memory by
two elementary experiments, the first to specify the
direction of current by a typical voltaic couple (zinc
and copper), the second to identify with such a current
the current of animal electricity that passes from more
active to less active tissue.

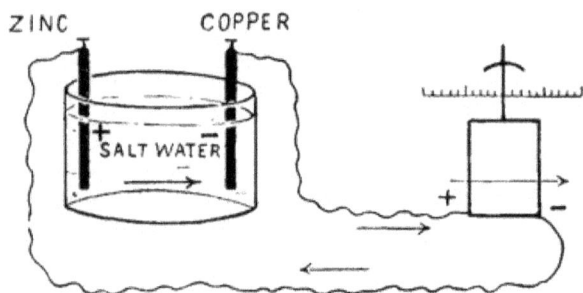

FIG. 2.—A simple voltaic couple.
   The surface between zinc and fluid is the principal seat of chemical action.
Current in the cell is from zinc through fluid to copper.   Through the galvan-
ometer it is from copper to zinc.

Preliminary Experiment.—A slip of zinc and a
slip of copper dipping into a glass of salt solution
and connected with a galvanometer will exhibit the
character and direction of action of a representative
voltaic couple.   The current will indeed be so large
that with this delicate instrument arranged as it is for
the far smaller currents generated in living nerve, we
must "shunt" it, *i.e.*, only let a small fraction of it
pass through the galvanometer.

The direction of deflection is as you see such as to indicate current in the voltaic cell from zinc to copper. Whatever you may choose to remember or forget, bear in mind that zinc is the active plate, at which chemical change takes place, and electromotive pressure takes origin. Later on when we come to deal with the electrical response of living nerve, we shall find it of the utmost convenience to make use of language based upon this fundamental notion, to speak of living nerve as being more or less zincative or zincable, as having greater or smaller zincativity and zincability.

Second Preliminary Experiment.—Let us in fact simplify matters further ; omitting the copper plate

FIG. 3.—A bit of zinc wire held in the right hand and touching one terminal of a galvanometer circuit (while the other hand rests on the other terminal) gives current in the body from right to left, in the galvanometer from left to right. The current is aroused at the contact between the zinc and the slightly moist skin, and is directed from zinc to fluid. (If the galvanometer wire, instead of ending at a metal terminal R, dips into a glass vessel of salt solution, and circuit is completed by dipping the zinc into the same solution, current would run the other way round —*i.e.*, anti-clockwise instead of clockwise—being as before from zinc to fluid.)

and substituting in the circuit our own body for the
vessel of salt water, by laying one hand upon one
terminal, and a zinc rod held in the other hand upon
the other terminal.

As before, you see that the current is from the zinc
through the body, and to the zinc through the galvanometer, i.e., the zinc is electro-positive.

You will find in this simple observation not only
a stepping-stone towards the proper understanding
of an animal current, but a ready practical device for
ascertaining in a complicated circuit what direction of
current at any part of the circuit is indicated by a
given deflection of the galvanometer.   If you put
your body into circuit by taking one end of a wire in
one hand L, and touch the other end with a bit of
zinc, you will get the deflection indicative of current
in your body from the zinc, i.e., from R to L in the
figure.   If that deflection is of the same direction as
that of the deflection under question, you know that
the latter was produced by activity on the R side ;
if the deflection is reversed, activity has been on
the opposite side.

Our first representative experiment is a realisation
of the imaginary case that we took as our type, and
had best be shown to you at least twice, on the two
representative excitable, that is to say, living tissues—
on a muscle namely, and on a nerve.

We will begin with muscle, with which the experiment is easier to make and easier to understand.

First Representative Experiment (on Muscle),—
This isolated muscle—it is the living muscle of a
dead frog—is connected with the galvanometer by
two electrodes touching it at two points T and L—its
tendinous end or transverse section, and its longitudinal
surface (fig. 4). As it stands, and though it has been
prepared as gently as possible, it is in all certainty
not absolutely uninjured and normal, therefore in all
certainty not homogeneous and electrically indifferent
at all points. It is living, that is to say it is dying—
for life is essentially slow death—and it is dying (or
if you prefer it living) in any case chemically changing,
more rapidly at its thin end T than at its thick part L.
If so—and for the purpose of making quite sure, I
may make it so by a touch of a hot wire—T will be
zincative to L, and we shall have current as shewn
by this diagram, illustrating on muscle that active
matter—in this instance stirred up by injury—is
zincative in relation to less active matter, giving
thereby a current of animal electricity.

This has been the first half of the experiment.
Its complementary half, illustrating the same principle,
will complete it and fix its meaning to you. The
nerve through which this muscle was controlled when
it was in its place in the body, has been carefully
dissected out and protected from drying so that it
is still alive. Its far end from the muscle is stirred
up by weak faradisation, and you see that the muscle
contracts—telling you by its contraction that it and

its nerve are still alive—and at the same time you see
that the spot of light deflected one way by the current
of injury from T to L, moves the other way (*i.e.*, undergoes a negative variation) when the muscle contracts.
This negative movement of the spot means that the
chemico-physical change taking place in the con-

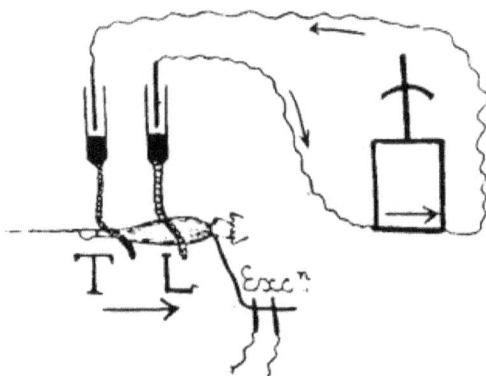

FIG. 4.—Muscle currents.
    The "current of injury" is from T to L in the muscle. The "current of
action" (negative variation of the current of injury) is from L to T in the
muscle.

tracting muscle is greater at L than at T, or that L
has become zincative to T.

    A little reflection will soon convince you that this
fact falls under our key principle. We started with
a current of injury due to a difference between T and
L ; T was active and zincative, therefore less capable
of further activity, *i.e.*, less zincable than L. When
by exciting its nerve, we stirred up the whole muscle,
T and L included, we obtained a greater augmenta-

tion of activity at L, the resting part, than at T, the already active part. The current of injury generated at T was due to a difference between T and L, its negative variation generated at L was due to a diminution of that difference. And there must be a difference before there can be a diminution of difference; there must be a current of injury for you to get a negative variation of it. If the muscle had been chemically homogeneous throughout, whether at rest or in action, with therefore any two points T L equally zincable or equally zincative, there could have been no (or little) difference aroused between T and L when both are equally (or nearly equally) altered.

Second Representative Experiment (on Nerve).— In the first experiment (on muscle) the nerve was used merely as an instrument to stir up the muscle, and the muscle spoke in two ways—by an actual movement of its substance, by a chemico-electrical change revealed in the galvanometer.

In this second experiment (on nerve) the nerve is to be used by itself, cut off from its natural organ of expression, which is muscle, and connected by the points T and L with a galvanometer that will serve as an artificial organ of expression.

We shall begin by testing the nerve for current of injury, expecting to find it directed in the nerve as in the muscle from disturbed Transverse section to undisturbed Longitudinal surface. And so it is.

Now, let us arouse the whole nerve—and for the

purpose of a distinct demonstration, I may hardly hope
to do so otherwise than by electrical stimulation—and
let us watch for a negative variation of our current
of injury. The variation is not large, but it is unmistakable ; (and while we are about it I will submit
it to two tests, the significance of which will be
explained later—the test of direction and the test of

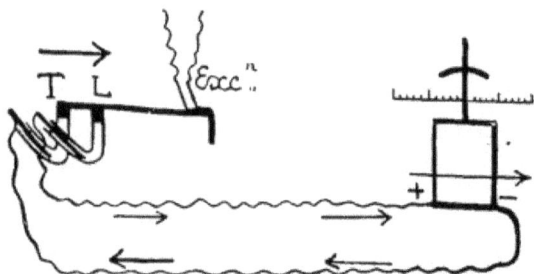

FIG. 5.—Nerve currents.
   The "current of injury" is from T to L in the nerve. The "current of
action" (negative variation of the current of injury) is from L to T n the
nerve.

distance. I have now reversed the direction of the
electrical stimulation, and the variation has remained
negative and of unaltered magnitude. I have now
brought the exciting electrodes much nearer to the
leading out electrodes, from a distance of about 3 cm.
to one of about 1 cm, and I repeat the test with both
directions of stimulating current ; still the negative
variation remains negative and of unaltered magnitude. We are thereby entitled to conclude that the
effect was a "true negative variation," and not due

to current escape nor to du Bois-Reymond's electrotonic current.)

The value at which you will be disposed to
estimate the electrical signs of physiological activity
will be greatly enhanced by the closer consideration
of the first experiment (on muscle). Here you have
in the mechanical effects of muscular contraction, an
obvious measure of its physiological activity—the
muscle shortens much or little according as its physiological activity is much or little, and one naturally
asks himself whether with this greater or smaller
mechanical effect there is a greater or smaller electrical effect.

A very simple variation of the experiment will
answer us in the affirmative, and the records of
more prolonged observations will confirm that affirmative beyond all doubt.

Let us go back to the experiment on muscle, and
make the muscle contract much or little, watching by
the galvanometer to see whether the size of the electrical effect corresponds. It is difficult to watch two
things at once, so in place of the muscle itself I
show you the end of a lever acted upon by the
muscle and marking against a smoked plate that
runs past the lantern ; the lever is the contraction
indicator, and the galvanometer is the current indicator. If now you will watch the galvanometer
spot while the muscle is stimulated more or less
strongly, you will see that the spot makes a greater

or smaller movement, and turning to the muscle
you will see that the muscle has contracted more
or less strongly.

This is a very simple but very important fact,
as a link in the argument that electrical changes
are a measure of physiological activity. More and
less muscular action involve more and less chemical
change, a larger and smaller diminution of the
current of injury. And whereas by arousing the
quiescent muscle you get a negative variation of
the current of injury, you would get a positive variation of the same by "quelling" an active muscle.
To be fully assured of this correspondence between
the mechanical and electrical effects of this one
cause — chemical activity, pray look at this double
record, where the white line gives series of contractions and the black line the corresponding series
of electrical effects of one and the same muscle
(fig. 6).

Each of the two series exhibits a declining effect—
the expression of muscular fatigue—and the rate of
decline is nearly the same in both cases. Not quite
the same, however ; if you examine the twin record
at all critically, there are evident divergences ; the
two records do not tell precisely the same story ; and
although the point is not one upon which we need
dwell now, I may observe that of the two versions—
the mechanical and the galvanometric—the latter is
the more accurate. It gives a declining series of

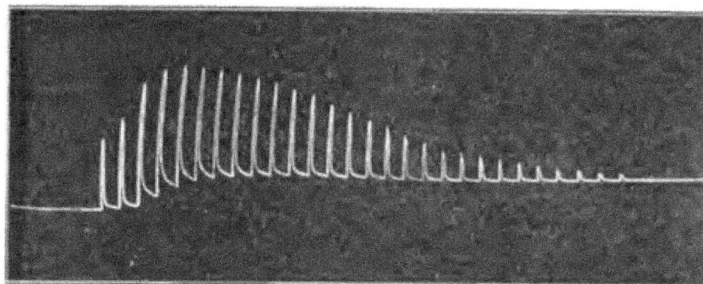

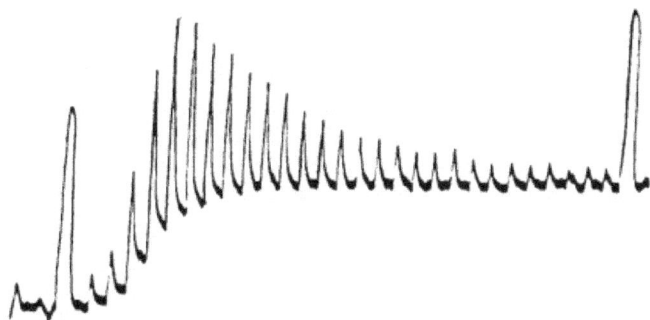

FIG. 6 (2462).—Simultaneous record of the mechanical and electrical responses of a gastrocnemius muscle (frog), excited by tetanising currents for $7\frac{1}{2}$ seconds (60 to 70 interruptions per sec.) at intervals of 1 minute.

The upper line (white on black) gives the series of lifts of a weight of 10 grms.

The lower line (black on white) gives the corresponding series of electrical effects ; the large deflections at beginning and end of the series are standard deflections made by $\frac{1}{100}$ volt. Dead-beat galvanometer (shunted).

In the first and second pairs of responses the strengths of stimulation were 4 and 5 units ; in the third and subsequent pairs the stimulation was at 10 units.

which the tops lie in a curve convex towards the base line ; and if I had adopted a rather better mechanical method (the so-called "isometric" method by which the muscle is made to record its tension against that of a stiff spring which allows the muscle to shorten but very little) you would have witnessed a declining series of tension effects much more like the galvanometer series.   But we shall come back to this point in dealing with the subject of fatigue ; for the present I am only concerned to establish in your opinion that the galvanometer is a good indicator of physiological modifications ; later I shall hope to convince you that it is in this respect the best and most accurate indicator at our disposal.

To watch the alterations of physiological activity taking place in fatigue or under other influences it matters little as regards the muscle itself whether you make use of a lever or of a galvanometer as your indicator.   Each record is almost a replica of the other, and contrary to your expectation perhaps, the less obvious method by galvanometer, once the preliminary difficulties have been overcome, is easier to work, as well as far more delicate and more accurate than the mechanical method.

More than one very significant inference arises from this simple observation, but I shall only comment upon one at this stage—to wit that the electrical effect is an exact measure of action in muscle, and may therefore be appealed to as a measure of action in nerve.

There is, however, one preliminary objection to the unlimited admission of this conclusion that must be taken at once.

We arouse action in the nerve by artificial "stimulation," and our artificial stimuli may be electrical, chemical, mechanical, &c. As a matter of convenience we most frequently employ electrical stimulation from an induction coil. May we be sure that the animal electric effects are not due to our electrical apparatus? An exhaustive examination at this stage of the means by which the answer is obtained would lead us too far afield; and I shall content myself with saying that while electrical effects aroused in nerve may sometimes be coarse fallacies, and are in general most pronounced when electrical stimulation has been used, yet they can certainly be produced by non-electrical stimuli, and even when produced by electrical stimuli, are phenomena proper to the *living* state of nerve. Reasons for admitting this will be offered presently by means of anæsthetics, as soon as we have clearly appreciated the bearing of this variation of our first experiment (on muscle), in which an electrical effect results from non-electrical excitation. The muscle is connected as usual with the galvanometer by points L and T, giving as usual the current of injury from T to L (in the muscle). I now excite the nerve by pinching it, *i.e.*, non-electrically; the muscle contracts and the current of injury is at the same time diminished.

That has been an electrical effect of a non-electrical excitation, and I might with a little care and a rather more sensitive galvanometer demonstrate a similar effect on the isolated nerve; finally, I might proceed a step further, and show that an electrical effect accompanies a natural discharge of nerve-impulses, as well as the impulses aroused by all artificial stimuli, whether these be or be not of electrical origin. But to do this would be superfluous to my present purpose, which is simply to show that an electrical stimulus is not indispensable to an electrical effect. Electrical stimulation is however so conveniently applied, can be so nicely timed and graduated, that except in a few rare cases we shall systematically make use of it, and it was therefore essential that you should realise at the outset that the electrical nature of the effect is not determined by the electrical nature of the stimulus. Later, when we shall examine the electrical effects manifested by the heart and by the retina, we shall see cause to be still more fully convinced of this.

In recollection of the fact that as regards muscle the mechanical and electrical response to stimulation run a nearly parallel course, and utilising the fact first ascertained by du Bois-Reymond that an impulse aroused at any part of a nerve is conducted in both directions from the stimulated point, we may establish a comparison between the changes

in a nerve and those in the attached muscle, stimulating the nerve in the middle of its course, using
as indicator of the state of the muscle its mechanical contraction, and as indicator of the state
of the nerve its electrical response. This has been
done in the experiment of which fig. 7 is the
record. The muscle has been attached to a lever
carrying a small paper flag that threw its shadow
through a slit upon a photographic plate, which at
the same time received the impression of a galvanometric spot of light moving in obedience to the
electrical changes in the nerve itself. The nerve
is stimulated once a minute; the muscle at one end
of the nerve contracts, the galvanometer at the
other end of the nerve exhibits a negative variation.
The pointed shadows of the flag moved by the
muscle, give on the developed plate a record of the
behaviour of the muscle; the excursions of the bright
spot give a record of the behaviour of the nerve.

Now the obvious points are these. By excitation
of one and the same nerve there has been a declining
series of muscular contractions and a uniform series
of negative variations of the nerve-current. The
muscle, to all appearance,[1] is fatigued, while the nerve
exhibits no sign of fatigue. It is thus very clear
that muscle does not faithfully express the state of

[1] As will be shown in a future lecture, it is more the nerve-
terminal (end-plate) than the muscle that is fatigued; but this
has nothing to do with the present argument.

2

the nerve itself, but only that of the combined
organ formed of the nerve, the nerve-terminal and the
muscle. It is equally clear that to follow the changes
taking place in the nerve, the galvanometer is preferable to the muscle. And, finally, if once we are

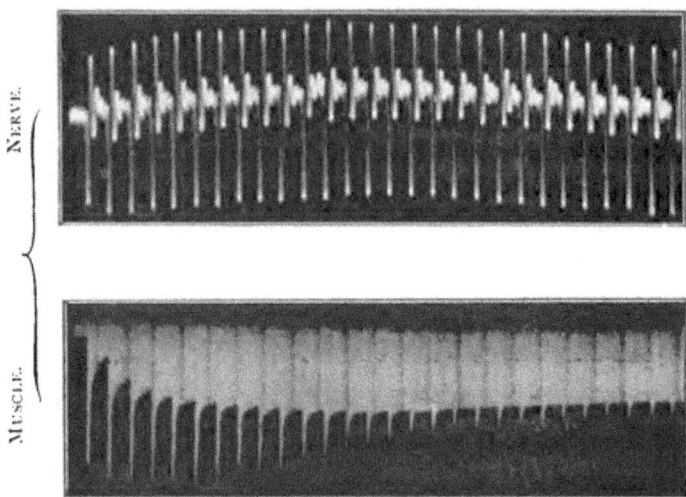

FIG. 7 (139).—Simultaneous record of the electrical response of nerve (upper
line), and of the mechanical response of muscle (lower line).
Stimulation at 10 units throughout. Ordinary galvanometer.

assured that the electrical indications are a true index
to the physiological state of nerve, regarded simply as
an excitable strand of protoplasm, it is evident that the
mathematical regularity of its electrical response to
regular electrical stimuli during long periods of time,
offers the best possible conditions and the best possible
object upon which to test the physiological modifica-

tions, effected or not effected, by all kinds of chemical
and physical interference.

But this very regularity of response has a suspi-
ciously non-physiological appearance. The first thing
to do is to make sure whether or no it is physiological,
and the first step to take is to test it by anæsthetic
vapours. The best and most expeditious anæsthetic
to employ for this purpose is ether (diethyl oxide).

Experiment.—The nerve resting upon two pairs
of unpolarisable electrodes, one pair for the exciting
current, the other serving to lead off the current of
injury and its negative variation to the galvanometer
(as shown in figs. 5 and 10), is enclosed in a glass
chamber into which air saturated with ether vapour
can be blown. The negative variation having been
provoked once or twice by completing the exciting
circuit, and having been found to be sufficiently large
and sufficiently regular, ether vapour is blown into the
nerve chamber. The negative variation is tested for
at intervals during the next two or three minutes; it
is seen to diminish and to disappear. This abolition
of the negative variation by ether is not permanent; at
the end of five or ten minutes the negative variation
reappears and gradually increases up to, and it may be,
beyond the normal.

This experiment (fig. 8), which represents the
regular and unfailing consequence of the etherisation
of a nerve, proves that the electrical effect we are
dealing with is a physiological phenomenon.

The first thought that suggests itself when one has seen by what a surprisingly regular series of electrical effects the nerve responds if regularly interrogated at regular intervals, and when one has satisfied oneself further that such electrical response is physiological, is that we are in possession of a most excellent test by which to recognise the action of chemical reagents upon the physiological properties of nerve.

For the test is applied to nerve and to nerve only. There is not, as when muscular contraction is used as the indicator, any question as to the share borne by muscle or by motor end plate. The experimental isolation of the object of observation is complete. It does not move. The excitation by which it is tested does not exhaust its excitability. It is a subsidiary, but by no means slight, advantage, that the photographic record of each and every observation can be easily taken, and preserved for future reference, and be as authoritative a century hence as it is to-day. The nerve has recorded its own series of answers during any treatment to which you may have chosen to submit it ; it cannot give a false answer ; the worst that may happen is that it may give some spurious answer to a spurious question, or that we may for the moment fail to correctly interpret the language in which its answers are returned.

A good instance to take in illustration will be to compare the action of different anæsthetics, and I will at once set going a chloroform observation on nerve in the manner now familiar to you, in order to allow time for the observation to be sufficiently cogent as regards one main fact in the subject, viz., that chloroform is far more powerfully toxic than ether, that its full effect is apt to be final, not followed by recovery within any reasonable lapse of time. Obviously this is a point of great practical significance, that may at any moment become a matter of urgent personal anxiety; and in what I shall have to say it will be impossible not to make statements having immediate practical applications. My experiments were not, however, primarily addressed to the practical issue; they were instituted solely for the purpose of testing the value of nerve towards its further utilisation as a test-object. And after all, anxious as I am not to infringe upon medical matters in presence of a non-medical audience, I shall prefer to explicitly formulate the inference as it presents itself to my mind, thinking on the non-medical lines of a purely scientific student, rather than to leave a perhaps hasty and popular inference to be drawn less guardedly and cautiously than might seem to be desirable.

This is the jubilee year—the jubilee day in fact—of the first trial of chloroform as an anæsthetic. On January 19, 1847, Sir James Simpson performed his first operation under chloroform anæsthesia. The manifestly beneficent effects of this powerful anæsthetic

led to its widespread preference over all prior and
subsequent rivals, so that in this country at least,
the designation "chloroformist" has almost become
synonymous with "anæsthetist," and many surgeons—
Lord Lister at their head—put all other anæsthetic
agents aside in favour of chloroform. Nevertheless,
like all potent drugs—potent because they are toxic—
perhaps above most potent drugs—chloroform, while
conferring quickly and effectually the great boon of
unconsciousness, exacts heavy toll. Not too heavy,
perhaps, if chloroform were the only anæsthetic in the
field, but far too heavy when we consider that there
are other, if less potent anæsthetics to hand, amply
adequate to the needs of nine-tenths of the cases for
which chloroform has been used with fatal results.[1]

It has been said that experimental physiology has
afforded no guidance in the use of anæsthetics—that
clinical experience is our only guide. I demur to this
statement, and shall at once exhibit to you a contrast
experiment representing, as in a nutshell, the comparative efficacy of these two principal re-agents.

---

[1] Dr. Lauder-Brunton, although pleading emphatically *in favour*
of chloroform as against ether, expresses himself as follows in his
Text-Book of Pharmacology (3rd Ed., 1893, p. 801):

"The operations in which death during chloroform chiefly
"occurs are short and comparatively slight, though painful, such
"as extraction of teeth, and evulsion of the toe-nail—operations
"in which the introduction of deep chloroform anæsthesia might
"be regarded as superfluous, and involving a waste of time."

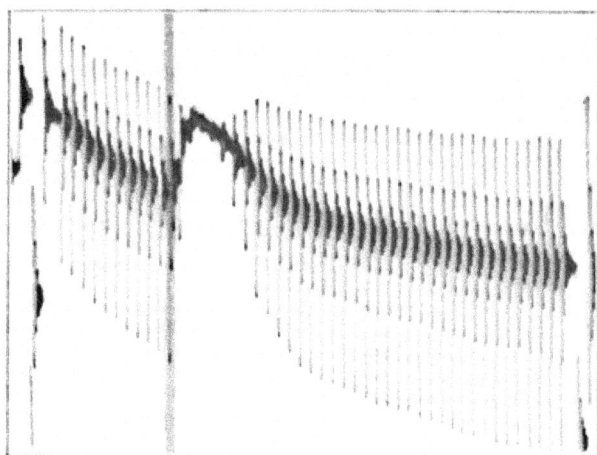

Before. **Ether.** After.

FIG. 8 (576).—Effect of ether (diethyl oxide, Et₂O) upon the electrical responses of isolated nerve. The electrical response is abolished for about five minutes; it then gradually recovers, and becomes somewhat larger than it was before etherisation. The excitations and responses of the nerve are at intervals of one minute.

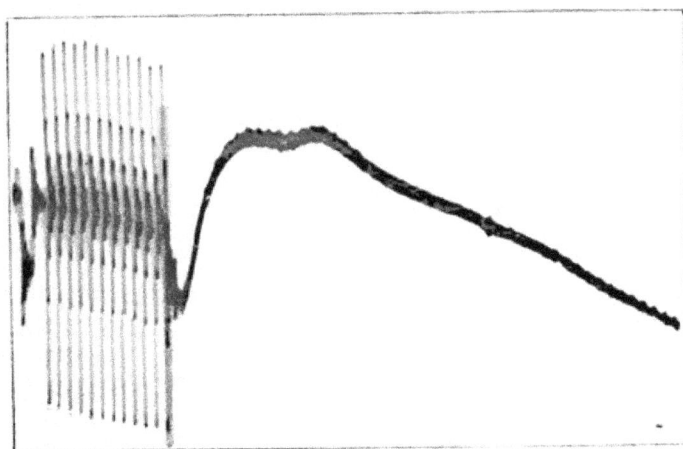

Before. **Chloroform.** After.

FIG. 9 (552).—Effect of chloroform (Trichlormethane, CHCl₃) upon the electrical responses of isolated nerve. The electrical response is definitively abolished; there is no recovery during the period of observation.

The action of ether you have just witnessed ; the effect was quite typical—an abolition of excitability consummated in about three minutes, and lasting another five minutes.

But how about chloroform ? I applied it in strong vapour for one minute nearly half-an-hour ago, and you then saw that the "excitability" was promptly abolished; on testing it now, you see that nothing happens; the nerve has definitively lost its excitability—you have witnessed its "death by chloroform."

Those are the facts in the rough—no slight difference of degree, but a striking contrast. There are many accessory considerations to be weighed ; obviously a question of this moment is not settled by a single experiment. How about quantity ? The nerve has had in each case a maximum dose, *i.e.*, for a period of one minute, air saturated with the drug, *i.e.*, with about 50 per cent. of ether and about 12 per cent. of chloroform. And to what extent may you argue from a naked tag of nerve to the complex human organisation? We shall not settle these matters offhand. We have seen that there is a similarity between the effects of anæsthetics on a living nerve and on a living organism ; our next task will be to examine some of the cases in which their effects differ.

HISTORICAL NOTE.

" Animal electricity " (*i.e.*, the electrical phenomena manifested by animals), is now a century old. In its biography the names of Galvani, Volta, Aldini, Matteucci, du Bois-Reymond, Hermann, and Bernstein, stand out as the principal marks, together with three somewhat bitter but fruitful controversies between Galvani and Volta (about 1794), between Matteucci and du Bois-Reymond (about 1850), and between du Bois-Reymond and Hermann (about 1867).

An admirable account of the Galvani *v.* Volta polemic is given by du Bois-Reymond in his " Untersuchungen über Thierische Elektricität " (Berlin, 1848-49-84), but the unfavourable verdict there passed upon the part played by Aldini in "these troubled waters," is, in my opinion, not borne out by the evidence.

" Animal electricity " may be considered as having taken origin from the observation by Luigi Galvani (or, according to some writers, by his wife Lucia), of spasms occurring, to all appearance spontaneously, in the legs of frogs prepared for the kitchen and suspended by copper ("copper" in 1791, "iron" in 1786) hooks to an iron railing. Did the act of discovery arise from Luigi or from Lucia ? Which of them on noticing the frog twitch first said "ecco!" might be interesting to know, but is of little moment. There are, however, certain points relating to the Galvani household that call for mention. At this period, Aldini, a nephew of Galvani, who was professor of Physics in 1793, was pursuing his studies in the University of Bologna, and was living in Galvani's house.

The memoirs bearing on the subject are :—

1. Galvani. De viribus electricitatis in motu musculari commentarius. Bologna, 1791.
2. Do., ristampate a Modena, 1792, con note e una dissertatione del Prof. G. Aldini (De animalis electricæ theoriæ).

3. Dell' uso e dell' attivita dell' arco conduttore nelle contrazioni dei muscoli. *Anonymous.* Bologna, 1794.
4. Do., Supplemento. Bologna ? 1794.

The two representative experiments are :—

   1. The contraction with metals.

   2. The contraction without metals.

The principal difference between Galvani and Volta was that according to the former the contraction with metals in the circuit, was due to animal electricity conducted by the metal, while according to the latter it was due to an excitation by ordinary electricity aroused by contact of dissimilar metals. The contraction with metals did not prove the existence of animal electricity. This could only be done by an experiment in which none but animal tissues were in circuit. Thus the origin of "Animal Electricity" becomes fixed at the "contraction without metals."

This experiment is first mentioned in the anonymous Trattato of 1794, not in Galvani's Commentary of 1791.

Who was the author of the anonymous "Trattato," and who therefore was the author of the "Contraction without metals"? Aldini in his "Essai sur le Galvanisme" alludes to the experiment as his own. But there is nothing to indicate whether Aldini or Galvani was the author of the Trattato.

On the other hand, Gerhardi in his annotations to the "Opere edite ed inedite del Professore Luigi Galvani," Bologna, 1841, and Du Bois-Reymond in his "Thierische Elektricität," unhesitatingly attribute the anonymous Trattato, including the important experiment, to Galvani alone as the "vero ed unico autore."

The manuscript of this Trattato, said by du Bois-Reymond to be preserved at the University Library of Bologna, and to be partly in Galvani's handwriting and partly in another's handwriting, is not now to be found. With regard to this, we are therefore thrown back upon probabilities, and these in my opinion, indicate *at least* a joint authorship of the Trattato, certainly not the exclusive authorship of Galvani, and certainly not a plagiarism by Aldini.

In 1794 (the date of its publication) Galvani was 57 years of age, professor of obstetrics and rector of the University—having previously held the chair of Anatomy, and published the following memoirs : de Ossibus 1762 ; de Renibus 1767 ; de Manzoliniana Supellectili 1777 ; de Volatilium aure 1783 ; and de Viribus Electricitatis in motu musculari Commentarius 1791.

While at the same date Aldini his nephew, who had long been a member of his household, was 32 years of age, and professor of physics in the University. In 1792 he published his Dissertation " De animalis electricæ theoriæ ortu atque incrementis." In 1796 it is he who demonstrates the new electrical experiments to Napoleon during his Italian campaign, and it is to Aldini, not to Galvani, that all Volta's controversial correspondence was addressed.

Matteucci's contributions to the subject are numerous and scattered—in Italian, French and English literature of the time, 1830-50. His " Essai sur les Phénomènes électriques des Animaux," Paris, 1840, placed by Müller in the hands of du Bois-Reymond, was the starting-point of the latter's life-long labours. This " Essai " was expanded to a " Traité des phénomènes electro - physiologiques des Animaux," Paris, 1844, and to his "Cours d'Electrophysiologie," Paris, 1858. Matteucci's permanent legacy to the subject is " the Second-ary Contraction " described by him in the Philosophical Trans-actions of the R. S. for 1845, under the title of " The induced contraction," and referred by du Bois-Reymond to an excita-tion of the secondary nerve by the negative variation of the primary muscle.

The differing points of view of the two men are thus given by du Bois-Reymond in the Preface of his " Thierische Elek-tricität " : " According to him [Matteucci] there are no elec-" trical currents in the nerve. The muscle-current circulates " in the muscle only after certain modifications due to pre-" paration, and is in no relation with its contraction. The " so-called nervous principle is to him from first to last a " special hypothetical mode of motion, which he chooses to

" represent under the figure of the ether vibrations, holding it,
" under all conditions, absolutely distinct from electricity.   I,
" on the contrary, am in favour of a diametrically opposite
" view.   If I do not greatly deceive myself, I have succeeded
" in realising in full actuality (albeit under a slightly different
" aspect) the hundred years' dream of physicists and physio-
" logists, to wit, the identity of the nervous principle with
" electricity."

Yet du Bois-Reymond was no vitalist; his animating
thought was to analyse the "vital forces" and to find their
physical and chemical components.   His endeavours con-
centrated themselves upon the study of living nerve and
muscle, of their electrical manifestations in particular.   He
laid, perhaps, undue stress upon these manifestations by
presenting them as evidence of an identity between the
nervous principle and electricity.   In his view a filament of
nerve was a string of dipolar electromotive molecules, positive at their middle and negative at their two ends; between
longitudinal surface and transverse end a current exists
(current of rest) which undergoes a diminution (negative
variation) during excitation of the nerve or of the muscle.
In the electrotonic state (Lecture V.) the molecules are re-
adjusted with negative poles towards the anode and positive
poles towards the kathode.

Hermann's presentation of matters differs from the above
in several particulars.   Electrical differences of potential do
not pre-exist in normal nerve or muscle, but are the consequence of injury and of physiological activity.   Injured
points are electronegative to normal points.   Physiologically
active points are electronegative to resting points.   Du Bois'
negative variation is an action-current.   Du Bois' electrotonic currents are the effect of electrolysis.

REFERENCES.

The controversy between du Bois-Reymond and Hermann is scattered through several memoirs. Du Bois-Reymond's case is given in his *Gesammelte Abhandlungen*, vol. ii., Leipzig, 1877, p. 319 (reprint of paper of 1867). Hermann's case is given in his *Handbuch der Physiologie*, vol. i., p. 235 (Leipzig, 1879).

A compromise between the two theories is proposed by Bernstein in *Untersuchungen aus dem Physiologischen Institut*, Halle, 1888, reproduced in Bernstein's *Lehrbuch der Physiologie* (1894), pp. 359, 449. The newer literature of the whole subject is given in Biedermann's *Elektrophysiologie*, Jena, 1895, English translation by F. A. Welby : Macmillan & Co. The theory of the "current of injury," or "demarcation current," was first given by Hermann in 1867, and is very fully described in his *Handbuch der Physiologie*, vol. i., p. 235.

The negative variation of the currents of muscle and of nerve was discovered by du Bois-Reymond in 1843 (*Thierische Elektricität*, vol. i., p. 425). It is now frequently referred to as the "*current of action*."

The literature of anæsthetics is a very extensive one. Among the best general accounts of the subject may be named Claude Bernard's *Leçons sur les Anæsthétiques et sur l'Asphyxie*, Paris, 1875, and Snow's "*Chloroform*" (edited by Richardson), London, 1858.

LECTURE II.

CONTENTS.

THE value of experimental results depends above all upon the method by which they may have been obtained, and the great end of any method is that it shall stringently insulate from amid a crowd of possibilities the particular phenomenon to be qualitatively observed, and if practicable, quantitatively measured. It is of the first importance that the method of inquiry should be precisely given, for that method gives measure of the degree of simplicity and therefore of cogency to which experimental analysis and criticism have been carried.

And short of this, even for the simple interchange of information, such as students of particular branches of science might desire from each other, it is necessary that the instrumental means by which each expert reads his little bit of natural knowledge should be clearly understood. Each province of knowledge has indeed its own language, instrumental as well as verbal, and it only happens too often that students of closely allied phenomena fail to exchange lights for no other reason than that they do not understand each other's jargon.

The first steps in a language are generally the dullest and most repulsive, but only by reason of those steps taken does any language bear its message to us. Where can be the interest of black marks on paper if we cannot read them, or of a spot of light wandering along a scale if we do not know what it is saying.

"Current," "Pressure," "Resistance," "Galvanometer," are some of the A B C labels that must have some meaning for us before we may be permitted to proceed a step further otherwise than as sham students.

Well, these labels having a meaning for us—I do not say their full and complete meaning—I should next ask you to make sure that we understand what is meant by saying that "current varies with pressure and inversely as resistance," and then being sure of my ground, I should invite you to follow me in the not very complicated details of the special case here figured.

The central object is the nerve, lying upon two pairs of electrodes, inside a glass chamber, through which gases and vapours can be driven. The nerve must not dry, therefore gases are made to pass in

through a wash-bottle half full of water. Through
the electrodes T and L the current of injury of
the otherwise undisturbed nerve passes to the keyboard. The other pair of electrodes (*Exc.*) are
connected with an induction coil ; a circular com-
mutator revolving once a minute sets up excitation
at minute intervals, and each time this happens there
is a negative variation of the current of injury. A
reverser enables the direction of the exciting currents
to be changed at will.

The keyboard is made up of four keys, K_1 connected with the nerve, K_3 with a galvanometer, K_2
with a compensator, K_4 with a second galvanometer.

When all four keys are closed, the board is simply
a square of metal. When K_1 is opened, a mouth is
made by which the nerve currents are let into the
square. When in addition K_3 is opened, a mouth is
made by which the nerve currents are let out of the
square to the galvanometer. On the same principle
when K_2 is opened, a standardising current is let into
the square, and thence through the nerve and the
galvanometer, and if need be the fourth key, K_4, is
connected with a second galvanometer : if there is no
such galvanometer in use, K_4 is left closed.

The galvanometer, by which your nerve currents
are measured, consists essentially of a suspended
system of little magnets fixed to the back of a light
mirror, which reflects a beam of light on to a scale.
The nerve currents traverse a coil of fine wire sur-

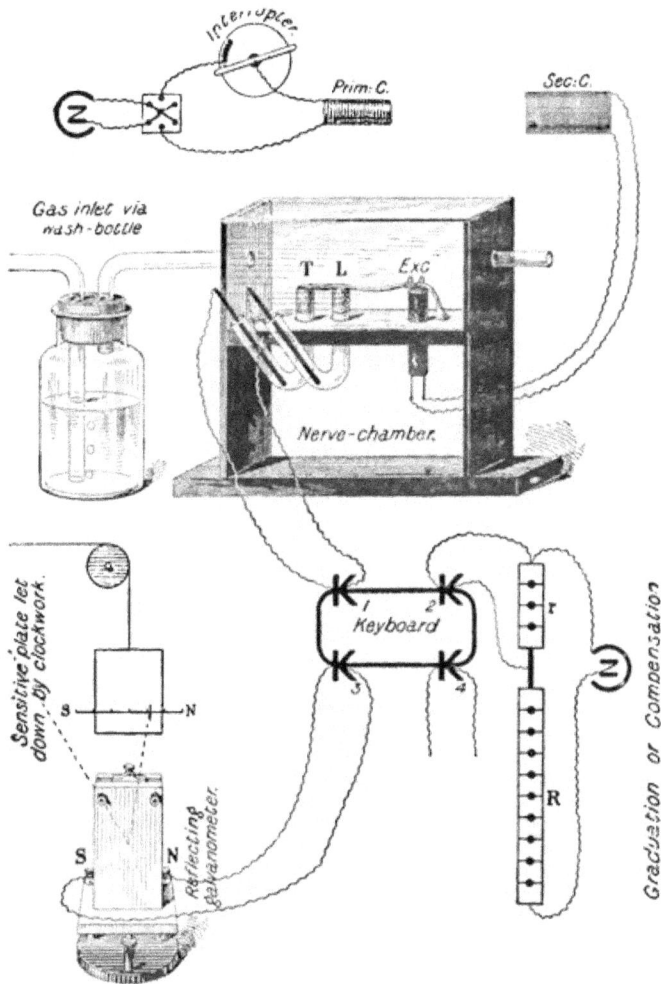

FIG 10.

PLAN OF APPARATUS.

(This figure is repeated on a fly-leaf at the end of the volume.)

rounding the magnets, deflect them, and with them
the mirror and beam of light which moves right
and left on the scale, serving as a pointer to the
amount of deflection, which is an indicator of the
amount of current. The light moves right and
left on the scale; replace the scale by a screen
with a horizontal slit, arrange a photographic plate
to descend vertically behind the screen, darken the
galvanometer room, and your pointer of light makes
its mark, and all its movements come out on the
developed plate as black lines on a clear background ;
you have a recording galvanometer.

It is a great convenience to have such an instrument by itself in a dark room well away from
coils and keys and moving people, and it is well
to leave the recorder severely to itself once adjusted
and started. But it is also very convenient to know
what is going on. This is effected by a second
galvanometer attached to K_4, upon which all that
is happening in the dark room can be watched on
a scale in the usual way. The resistance in circuit
added by the second galvanometer is of no moment,
that of the bit of nerve between T and L is already
100,000 ohms, that of the second galvanometer is
perhaps 10,000.

In the figure the galvanometer (and mirror) are
as if viewed from the back, the sensitive plate faces
you. Imagine K_1 and K_3 open, so that the nerve
current passes through the galvanometer, and trace

the current from T to L, you will find that it will
be from S to N through the galvanometer, like-
wise that the negative variation from L to T will
be from N to S, and when we see the record that
I am about to take during the next half hour, you
will realise as convenient and natural that the plate
should be turned to show deflections up and down
instead of right and left. The negative variation
then reads downwards, the current to which it is
negative reads upwards.

I have not yet noticed the apparatus connected
with K. for "Graduation or Compensation," but
you will readily appreciate its use if you know that
current increases as pressure and inversely as resistance.

Take a case like this : you have made an experiment with carbon dioxide, let us say, and on measuring out the record you have a negative deflection of
10 before and 20 after carbon dioxide. The current
has doubled, but this might be by doubled pressure
or by halved resistance, or from a mixture of the
two changes. You do not know where you are, but
you would do so at once if you had as terms of comparison the deflections made by a convenient standard
pressure before and after carbon dioxide. If they
were found equal, then the resistance is known to be
unaltered, and the augmented deflection was due to
pressure only. And from any inequality you would
be able to correct for resistance. If the standard

deflection had fallen from 10 to 9 then your deflection 20 indicates a pressure of 20 × $\frac{10}{9}$, if it had risen from 10 to 12, then the deflection 20 indicates pressure of 20 × $\frac{10}{12}$.

For nerve, it is convenient to take as a standard the deflection by $\frac{1}{1000}$ volt (for muscle we shall take it by $\frac{1}{100}$ volt, for the retina by $\frac{1}{10000}$ volt). Say that the pressure of the cell is 1 volt, this will be done by taking two resistances in the proportion of 1 and 999 (large enough at any rate to let us neglect the internal resistance of the cell), connecting the cell with the two extremes, and the key K_2 with the two ends of the smaller resistance. This is because the difference of pressure at the two extremes being 1 volt, and falling progressively from beginning to end in proportion with resistance, if we take a given resistance between any two points (say 1 ohm), we shall have a difference of pressure between those two points equal to $\frac{1}{1+999}$ or $\frac{1}{1000}$ of the total between the two extremes

For all our purposes it will be sufficient to take a Leclanché cell (assumed as 1·45 volt) and fixed resistances (1 and 1450 ohms) to deliver $\frac{1}{1000}$ volt. And as a matter of practical convenience, which will be found a convenient ear-mark for reading our plates, we shall put up connections so as always to take this standard deflection in one direction, viz., towards the left or S or negative side of the scale, i.e., downwards on the completed record.

The usual way of measuring pressure or potential is,

however, by compensation, and you may, if desired, use the resistance boxes r R for this purpose. You have only to see that the current from r opposes the nerve current at the keyboard, and by successive trials find that value of $\frac{r}{r+R}$ at which there is no current through the galvanometer. The nerve current is then compensated or balanced just as a given mass in one scale-pan is balanced by an equal weight in another scale-pan. Say, e.g., that the nerve current is balanced when $\frac{r}{r+R}$ is $\frac{10}{1000}$, then—taking the pressure of the Leclanché cell at 1·5 volt—the pressure of your nerve current is 0·015 volt.

It has been customary to take observations of the excitatory changes (negative variation, &c.) of nerve, with such compensation in force, and for rheotome observations this is necessary. But for observations such as these it is a mistake to use a compensating current ; in its presence we have opposing each other in the nerve (a) the injury current from T to L ; (b) the compensator current from L to T, and the latter current has its anode at L. We shall find later that a foreign (polarising) current increases the zincability of living nerve at its point of entrance or anode.[1] The negative deflection due to excitation and zincativity aroused at L is therefore unduly exaggerated— reinforced by what we shall recognise as the polarisation increment of the compensating current.

---

[1] See below, the polarisation increment, p. 135.

We shall therefore neglect compensation, and if the nerve current is so large as to send the spot " off scale " or " off plate," readjust the deflection by means of a controlling magnet such as that fixed above the case of Thomson (Kelvin) galvanometers.

We shall have yet another use to make of the resistances r R when we come to study the polarisation increment. We shall then require to deliver to the nerve, pressures of, say, 0·1, 0·2, 0·3, &c., volt, and shall do this without altering the connections, by adjustment of the resistances.

I will illustrate the use of these resistances, and at the same time give an example of the principle that deflection (by current) varies directly as pressure, and inversely as resistance, by the following pair of experiments. This is the circuit (fig. 11); it need

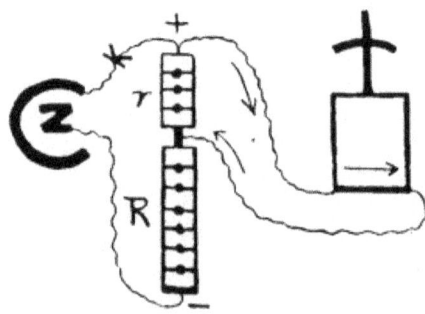

FIG. 11.—A Compensator.

not be separately put up; it is sufficient to close the superfluous keys K_1 and K_4 of the circuit as given in fig. 10, leaving only K_1 and K_2 open.

I will take r small and R large, so as to take R as a sufficiently accurate denominator of the fraction $\frac{r}{r + R}$. Taking R at, say, 100,000 ohms., and r successively at 1, 2, 3, 4 ohms, you see that the deflections are 1, 2, 3, 4, i.e., current varies directly as pressure. Taking r at, say, 4 ohms, and R at 100,000. 200,000, 300,000. 400,000, the deflections are 4, 2, $1\frac{1}{3}$, 1, i.e., as 1, $\frac{1}{2}$, $\frac{1}{3}$, $\frac{1}{4}$, i.e., current varies inversely as resistance, which you may recognise as an instance of the fundamental principle known as Ohm's Law.

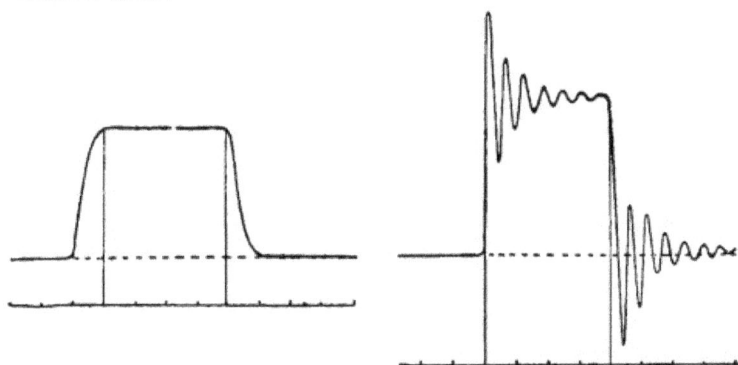

FIG. 12.—Records of deflections by the dead-beat and by the ordinary (partially damped) galvanometer.

The time-records are in quarter-minutes; the "falling time" of the dead-beat magnet is about 15 seconds; the partially-damped magnet has 7 complete oscillations in 1 minute, i.e., an oscillation period of $8\frac{1}{2}$ seconds.

In the galvanometer now at work before you the movements of the magnet are partially damped; you see that at each deflection the spot swings to and fro, above and below its position of rest, by a series of diminishing oscillations. Many of the records you

have seen have been taken with a galvanometer of
this character; others have been taken with a fully-
damped or dead-beat instrument, in which the magnet
comes slowly to its position of rest without overshoot
or oscillations.

Each instrument has its advantages and dis-
advantages. The former is more sensitive, and
best adapted to show effects of short duration,
the latter to show more prolonged effects uncom-
plicated by oscillation; and for after-effects in the
nerve they supplement each other. The time of
an oscillation to and fro, or "period" of the partially-
damped magnet is, say, eight seconds, and the range
of a swing is, say, double that of the next swing, *i.e.*,
the "decrement" is 2. A swing, caused by an
electrical change lasting four seconds and ending
abruptly, will be followed by an instrumental after-
deflection in the opposite direction of, say, ⅖,
or about 0·7 of the original swing. And if the
change, instead of ending abruptly, either goes on
of the same character, or gives way to an opposed
after-change, we shall have the after-oscillation either
checked and reduced, or helped on and augmented.
If this reduction or augmentation is very pronounced,
we shall be justified in saying that within the four
seconds immediately after an excitation has ceased,
the after-state of the nerve has been like or unlike
its state during the excitation. It would not do to
lay too much stress on slight alterations, but coarse

alterations such as those exemplified in figs. 26 and 27 are not ambiguous.

With a dead-beat magnet, which moves sluggishly, as if plunged in a sticky fluid, matters are quite different. The time it takes to pass from one position of rest to another is its "rising" or "falling" time. Any brief impulse that it may receive is lost; and while it is falling back to a position of rest at the termination of an effect, it will not show whether the after-state is positive or negative to the previous state. It is only if the after-state is very prolonged, in comparison with the falling time, fifteen seconds, that the magnet will indicate its presence, and then we practically only get signs of the after-state from the end of the falling time onwards.

Thus, however, the two kinds of magnet supplement each other as witnesses; both indicate the state during excitation, the swinging magnet indicates the state immediately after excitation, the dead-beat magnet the state during a later period after excitation.

Let us now turn to perhaps more familiar matters, and make use of this apparatus to test the effects upon nerve, of alcohol, soda water and tobacco smoke.

A nerve is lying upon its electrodes in the moist chamber (fig. 10). The wash-bottle is half full of fresh soda water, all the keys are closed, the galvanometer spot is at zero. I open K_1 (from the nerve) and K_3 (to the galvanometer) and the spot moves half across the

scale by reason of the current of injury from T to L.
in the nerve. What is the value of the pressure
difference between T and L giving that current?
I open K_2 from the graduation coils, and find that
the $\frac{1}{1000}$ volt gives 1 degree deflection, therefore the
current of injury that gives 8 degrees, is by a pressure of $\frac{8}{1000}$ volt. Now let us start the excitation by
which the nerve is questioned, and watch the negative
deflection that is its answer. It is a fairly good one—
about 1½ degree, i.e., 0·0015 volt. Now for the soda
water. I am passing air through it into the nerve-
chamber for one minute. The first deflection after
this is not much altered, but a minute later it is
distinctly larger, and at the third minute it is about
doubled. Is this greater deflection due to raised
pressure or lowered resistance? The standard pres-
sure of $\frac{1}{1000}$ volt let into the circuit by opening K_2 is
unaltered; therefore the augmented deflection is due
to augmented pressure, not to diminished resistance.

We go through the same steps with tobacco smoke
on a fresh nerve, and the result is precisely similar—
even more marked than in the previous experiment.
Like soda water, tobacco smoke is a stimulant.

And now to bring these trivial experiments to a
close, we will try the effect of strong alcohol, driving
its vapour as before into the nerve-chamber. The
response of the nerve is promptly abolished. The
nerve seems to be dead, but is in reality no more
than " dead-drunk "; I have little doubt of its

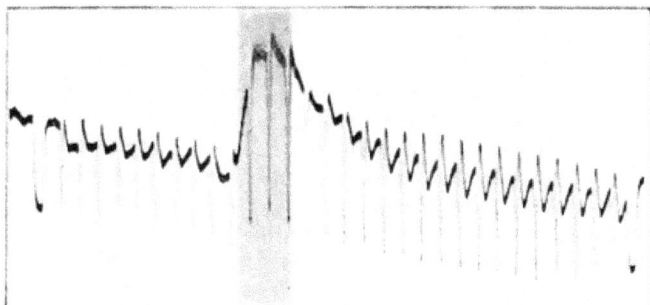

FIG. 13 (2466).—Effect of "soda water" upon the electrical responses of isolated nerve.

The vertical dark bar indicates when the "soda water" acts on the nerve.

The deflections at beginning and end of the series are standard deflections by $\frac{1}{1000}$ volt.

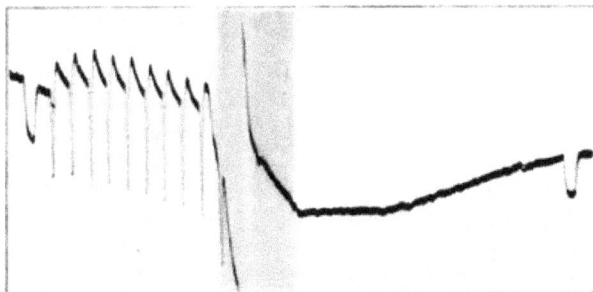

FIG. 14 (2464).—Effect of alcohol vapour (ethylic alcohol, EtOH).

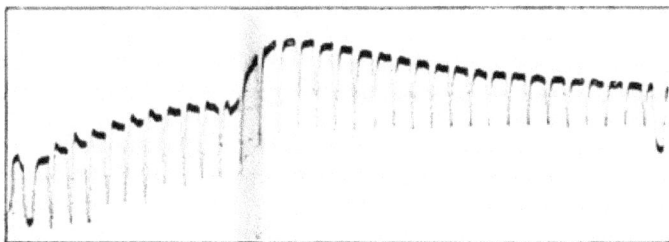

FIG. 15 (2468).—Effect of tobacco smoke on the same nerve (2464) twelve hours later.

recovery within a few hours, almost certainly by to-morrow morning.

Let us pause now to consider the effects of these three experiments, which are chosen on purpose to enforce the principle insisted upon at my last lecture, that living matter in the shape of a nerve, may give us some clue to the action of drugs on the living body. The soda water has acted evidently by reason of its dissolved gas, which is carbon dioxide. The tobacco smoke has very likely acted by reason of the same constituent, carbon dioxide ; still I am not prepared to say that no other constituent of tobacco smoke is active, since I have not yet tried the effect of tobacco smoke deprived of its CO_2.

Alcohol (that is to say ethylic alcohol, EtOH ; we shall on some future occasion have something to say about other alcohols) has, in the way we used it, acted as a profound depressant. If we had been very careful, we might have witnessed on the nerve its preliminary exhilarant effect. As to how it acts, I do not wish to be positive, but may remark that it not improbably acts by depriving the nerve of water — the effects of alcohol vapour and of drying are very much alike, and so indeed is that of prussic acid—a point that might perhaps be very effectively made use of by a temperance advocate. But such a person would probably omit to comment upon the effect of pure water, which very promptly puts an end to the nerve. Here, however, we come to a point of divergence between the bit of nerve and the entire organism; pure water has certainly no such toxic effect on man as it has when directly applied to one of his tissues.

Of the several materials just alluded to, by far the most interesting and important is the gas, carbon dioxide. It has perhaps been a little unfortunate that I fell into the hands of carbon dioxide so early in the investigation, for its allurements have been so great, the hints it has vouchsafed have been so significant yet so mysterious, that it has crushed out nearly all other claimants upon my time, and prevented me from pursuing as extensively as I could wish that superficial survey of a large number of chemical reagents which in the infancy period of an inquiry is the necessary preliminary to further and deeper analysis of the mode of action of a few chosen substances.

Nevertheless, considering that carbonic acid is one of the two principal disintegration products of all living matter—the vehicle of the respiratory export of carbon from the organism—that it produces with unmistakeable clearness effects that might have been predicted *a priori* were we only endowed with sufficient imaginative wit; considering further, that from these data we shall be able to draw inferences as to the existence and nature of physical changes effected within the nerve itself, and put those inferences to the test of clear-cut experiment, I cannot regard it as matter for regret that so much time should have been necessary to investigate carbonic acid at all thoroughly. And after all two years is not so very long, but the investigation is not near its end.

Let me at once, before entering upon any close analysis, exhibit to you two observations that represent the regular and typical action of carbonic acid upon nerve. The first of these two figures shows that carbonic acid in small quantity acts as a "stimulant." The next figure shows that carbonic acid in large quantity acts as a narcotic. It is in fact a typical anæsthetic—excitant when its action is slight (fig. 16), depressant when its action is complete (fig. 17).

It will not be amiss if at this stage I invite you to make a brief digression in order to recognise the relations of carbon dioxide in the economy of living matter, and perhaps broaden the view you may have formed of the respiratory function. It happened to me a few years ago to overhear a conversation in a conservatory between a young gentleman and a young lady. They were discussing the plants, and the two disputants (they were of quite tender years, nine and eight respectively) took up what struck me as very typical standpoints. The man-child said "of course plants breathe, how could they live if they didn't breathe"; the woman-child said "of course plants don't breathe, how can they breathe without lungs." Well, of course—

Before.   CO₂.   After.

FIG. 16 (674).—Effect of " little " CO₂.   Primary Excitation.

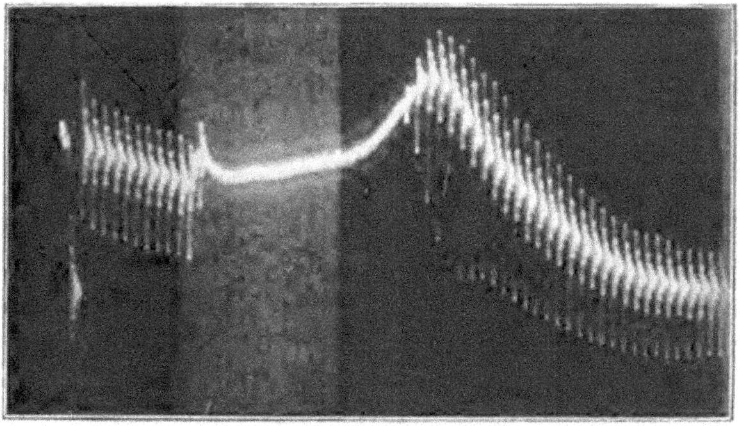

CO₂   After.

FIG. 17 (627).—Effect of " much " CO₂.   Complete Anæsthesia followed by Secondary Excitation.

the woman got the best of the argument, but the
man was right, and gave evidence, I think, of
superior imaginative power. We breathe by lungs,
so do all our personal friends among animals; but
fishes breathe by gills, plants breathe by leaves,
and protoplasm, devoid though it be of diverse parts
or organs, breathes by direct take and give between
itself and the atmosphere. And in an organism
where the living units are buried away from the
atmosphere, lungs or gills, heart vessels and blood
are the intermediate instruments of this take and
give between protoplasm and air. Protoplasm, then,
directly or indirectly, takes from the air oxygen, and
gives to the air carbon dioxide, and such respiration is the primitive essential function to which the
instrumental means — organs of respiration, organs
of circulation are secondary and accessory complications.

It has been said that "respiration is a slow combustion," and Black, a century and a half ago, fifty
years before the discovery of oxygen by Priestley, and
the closer identification by the Lavoisier school of
respiration with combustion, showed that the end-
product of respiration is identical with that of combustion. And even to-day, if we bear in mind that
the carbon dioxide produced by respiration (as indeed
that produced by combustion) is not a product of
direct oxidation—that carbon and oxygen do not, so to
speak, meet and join hands directly, either in the lungs,

or in the blood, or in the tissues, and straightway run
down hill together—I think the phrase "respiration
is a slow combustion" is permissible. Oxygen, the
incoming "food"—for oxygen is even a more immediately necessary "food" than water, bread and meat—
first climbs a mountain where it is lost to view, as a
member of that mysterious company of atoms we call
"proteid,"[1] which have power to act, which by acting
are expended, yielding to the body movements that
are the expression of its life, heat which is a *sine quâ
non* of its action, and carbon dioxide which is the
token and measure of its activity.

The immediate antecedent of physiological activity
is chemical change—a disintegration of "inogen," of
which at least one end-product is carbon dioxide.
Living tissue is deoxydative or reducing towards the
oxygen-yielding blood by which it is traversed; it is
at the same time oxydative and synthetic as regards
the formation of its own working balance of inogen,
deoxydative and analytic as regards expenditure of
that balance. It is a maxim of political economy that
imports equal exports; a similar maxim applies to the
organism and to any portion of the organism that is to
maintain itself in the physiological economy. If a

---

[1] Perhaps it would be better to say "inogen," in order not
to seem to prejudge the vexed question whether the force-
producing compound that disintegrates is proteid or carbo
hydrate.

4

muscle works and does not diminish, carbon and oxygen must come to it, as carbon with oxygen leave it. There must be integration reparative of disintegration. And although no doubt during heightened action the disintegrative movement is hastened, and during subsequent repose reintegration, yet we must not too absolutely imagine as independent and dissociated these two opposite movements—they are rather the concomitant aspects of *one* complex process of metabolism, although, no doubt, action gives prominence to the negative movement of expenditure, and rest favours the positive movement of recuperation.

Turning back to the electrical signs of nerve action, we find this opposition of movements mirrored as it were by the currents during and after action. During action we have a negative effect, after action we have a positive after-effect. To which double statement I may add that in certain states of nerve we may witness even during action a positive effect that may possibly signify that even during action integration may predominate over disintegration. But however that may be, we shall find in the empirical study of reagents that this positive variation contributes some very significant items to the series of arguments upon which we are about to enter.

To sum up, we shall have as elements of study three principal effects upon a nerve set up for experiment as shewn in fig. 10.

(1) The negative variation of du Bois-Reymond, due to disintegrative activity at L.

(2) The positive after-variation of Hering, due to reintegrative chemical movement at L.

(3) The positive variation to which I have alluded above, present only in certain states of nerve, attributable to integrative chemical movement during augmented activity at L.[1]

In the first case zincativity at L, deoxidation of inogen, rejection of CO_2, acidificatory effect.

In the second, and possibly third cases antizincativity at L, oxidation of inogen, deacidification.

But I cannot interpret matters at this stage with any degree of assurance, and shall therefore refrain from the attempt. Let me rather emphasise the bare fact that nerve is an extraordinarily sensitive reagent to the presence of CO_2. Here is a fresh nerve, giving a by no means impressive response, but I breathe through the little vessel that contains it, and there will be a perfectly obvious improvement of the response within one or two minutes, due to the carbonic acid contained in expired air. The nerve is by far the most sensitive reagent I know of to carbonic acid ; left for a minute or two in a tube of one cubic centimetre

---

[1] The conceivable changes at T are left out of count ; there is, however, evidence to support the view that the negative variation is often the sum of two homodromous activities— zincativity at L and anti-zincativity at T.

capacity containing 1 per cent. of carbonic acid, it is unmistakeably altered. Thus it has reacted to a hundredth of a cubic centimetre (or about one fiftieth of a milligramme) of CO_2, and no doubt this has not been a minimum limit.

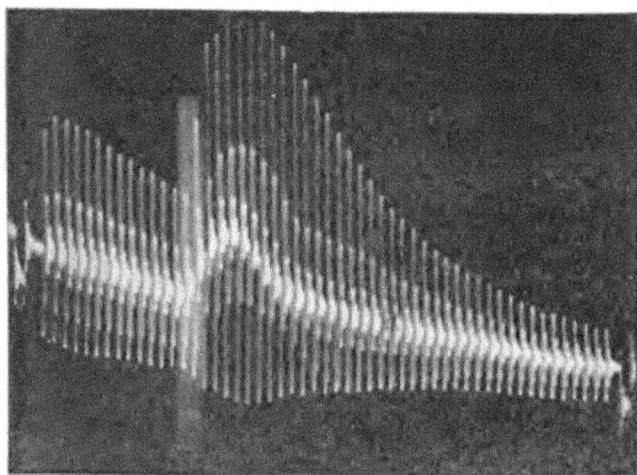

FIG. 18 (598).—Effect of expired air (Excitation by "little" CO_2).

In conclusion let me place before you the record of a complete experiment such as that of which you have just witnessed an initial stage. Plate 589 is an instance of the effect of expired air in an observation lasting about forty minutes. The series of re-

sponses to the left of the light bar gives the normal
of this particular nerve—the light bar shows when
expired air was blown in—and the next twenty or
thirty responses give token of the altered state of
the nerve during the next twenty or thirty minutes.
The expired air contained 3 per cent. of CO_2 and was
blown through the nerve chamber for two minutes.

By other experiments it was ascertained that the
alteration was neither due to altered temperature nor
to anything else in the expired air but CO_2. It is
then evident that nerve is extraordinarily sensitive
to CO_2, and this clear fact will form the point of
departure of our further study.

REFERENCES.

The general notion of " internal respiration," the functional
attribute of all protoplasm, as distinguished from " external respiration," the accessory means, alluded to on
p. 48, is developed by Claude Bernard in his *Leçons
sur les Phénomènes de la Vie communs aux animaux et
aux végétaux*, vol. ii., Paris, 1879. The notion of the
double aspect of functional changes in living matter,
expressed in the terms—analysis and synthesis, breaking-
down and building-up, dissimilation and assimilation,

katabolism and anabolism, disintegration and reintegration—is developed at length by Claude Bernard in *Phénomènes de la Vie*, vol. ii., by Hering in "*Lotos*," ix., 1888,
Zur Theorie der Vorgänge in der lebendigen Substanz, and
from the particular standpoint of electrical phenomena by
Hering's pupil Biedermann, in several papers summarised
in his *Elektrophysiologie*.   Hering's paper of 1888 is
translated in the current volume (1897) of "*Brain*."

LECTURE III.

It was comparatively early in these experiments— during the month of October, 1895—that the nerves under study presented some puzzling features in their mode of response, that for some time were a grievance and a stumbling-block. The nerves said "check," just as this month (February) they are again saying "check." My disgust at their enigmatical conduct came to its climax on October 27 with plate 672, and vanished in the evening of the same day with the appearance of plate 675.

I will venture to tell the story of these two plates, for I think that it illustrates what so often happens in the course of an inquiry as almost to be the rule, viz., that "check" in the course of some otherwise smooth and glib interrogation, although at first disliked as a stumbling-block, ought in reality to be welcomed as a stepping-stone in disguise, a sign that some unexpected bit of truth has cropped up.

I had undertaken to show a friend the stimulant action of carbon dioxide, and had announced it as an unfailing experiment. I had previously noticed that if a frog is killed and its two sciatic nerves used in succession, the second nerve always gave larger variations than the first. I had also noticed that in the case of a nerve that had been left lying for twelve hours in the killed animal, the variation was extraordinarily large—four or five times its usual value. But I had never tried the effect of carbonic acid upon such a variation.

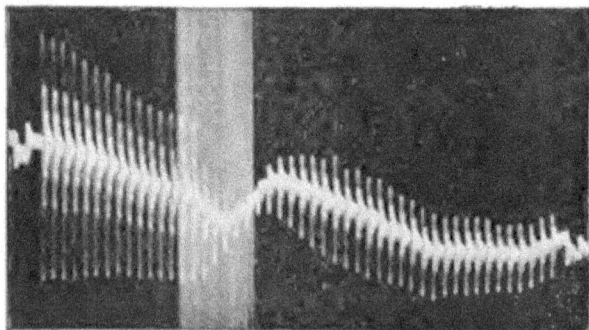

FIG. 19 (672).—Effect of CO_2 upon a "carbonised" nerve, i.e., a nerve left for 24 hours subject to the reducing action of the surrounding muscles.

Wanting to make the demonstration as clear as possible with a good variation to start with, I took a nerve that had been left for twelve hours in the frog and got the result recorded on plate 672 (fig. 19), to my considerable disgust. The stimulant action of carbon dioxide was conspicuously poor, and I don't think my friend watched the experiment out, in fact,

I recollect that the experiment of which you saw the record last week (fig. 16, plate 674) was made upon a fresh nerve immediately after his departure.

Later in the day the failure (which was the first I had met with) came under discussion. Clearly this nerve had been very much altered in its disposition towards carbonic acid. A reducing action of the surrounding muscles must have been at work : CO_2 of course. But has this been only muscle CO_2 or also nerve CO_2? Who shall decide and how? That question must wait, but the obvious question whether exaggerated activity of nerve is or is not attended with an increased production of CO_2 can be tested at once, since the nerve itself is such a delicate reagent to its presence from outside. A little expired air, acting on the nerve by reason of a quantity of CO_2 much below what can be detected by orthodox chemical tests, produces a great change of the galvanometric response (see fig. 18) ; surely if any change takes place within tetanised nerve, if one single molecule of CO_2 is produced within it, we shall have the characteristic alteration of response. Let us do this : take say five normal responses, then tetanise the nerve continuously for five minutes, then take some more responses. If one single molecule of CO_2 is evolved within the nerve, we shall have an augmented response, diminishing minute by minute with the disappearance of CO_2.

A blackboard sketch (fig. 20) accompanied this rather extravagantly expressed argument, and the slope

of the chalk line, indicating the state of things during
the proposed five minutes' tetanus, was dictated by the
thought that with the imagined progressive evolution of
CO_2 there should be a gradually increasing negative
variation, giving a gradually increasing departure from
the general base line.   Such was the forecast—yet,

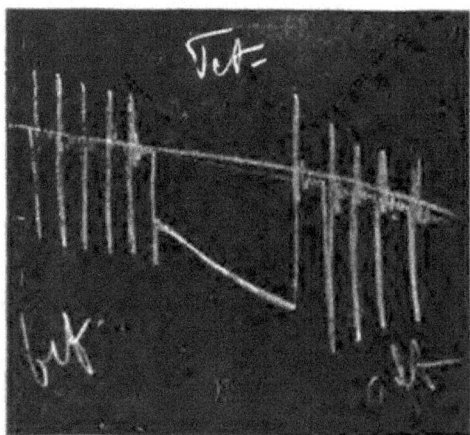

FIG. 20.—Hypothesis.

when half an hour later, I watched the lines of plate
675 (fig. 21) appearing in the developing dish, it was
with something of astonishment that I found the greater
regularity of the series to be the sole difference
between the autographic record of the nerve itself,
and my rough blackboard drawing.

Well, this is not proof, but it is very remarkable
evidence.   And considering that we have no previous

evidence of any chemical or physical change in
tetanised nerve, it seems to me not worth while
pausing to deal with the criticism that it is not CO_2.
but "something else" that has given the result.

Moreover, the evidence is corroborated in a very
remarkable manner under a considerable variety of
conditions.

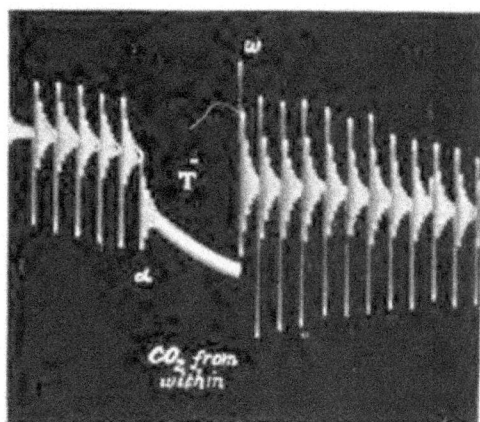

FIG. 21 (675).—Verification.

But before we are in a position to properly
appreciate this further evidence, it will be necessary to
enter into a digression, speculative from one point of
view, empirical, however, as regards our experimental
survey of the effects of carbonic acid acting from
without and presumably acting from within.

The typical electrical effect of excitation of a
freshly excised nerve is the negative variation of du

Bois-Reymond, and the hardly less typical after-effect
is the positive after - variation of Hering. The
electrical state of nerve after excitation is for a short
period the reverse of that obtaining during excitation ;
and if we call the latter " negative," we must call the
former "positive." Here, for example, recorded by a
quite dead-beat galvanometer, is a typical series of
negative followed by positive after-effects, the former

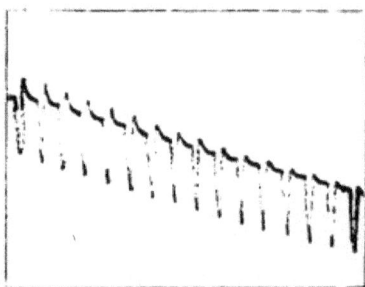

Fig. 22 (2177).

undiminishing, the latter progressively diminishing.
That progressive diminution of the positive after-effect
will presently be clear to us as a characteristic change
produced by carbonic acid—but I anticipate.

The positive after-effect presents itself more or less
obviously—more so as a rule in nerve that is becoming
a little stale, or that has been trifled with by previous
experiments. Sometimes the positive after-effect is so
pronounced as greatly to exceed the negative effect
that precedes it. As a matter of convenience I
distinguish such nerve as being in the second stage,

neither undoubtedly fresh nor unmistakeably stale—
transitional in short.

Finally it is not uncommon—nor yet common—to
witness, even during excitation, what I have been led
to regard as characteristic of stale or experimentally
maltreated nerve, viz., a positive electrical effect, which
has generally prolonged itself in more or less pronounced degree as a positive after-effect subsequent to
excitation, and has rarely given place to a negative
state during the first few seconds after excitation. I
lay, however, no stress upon this last point ; indeed, it
is chiefly for convenience of description that I make
this, for the moment empirical, division into three
stages—fresh, transitional, and stale, of which the three
characteristic features are respectively—(1) a negative
effect, (2) a positive after-effect, (3) a positive effect.[1]

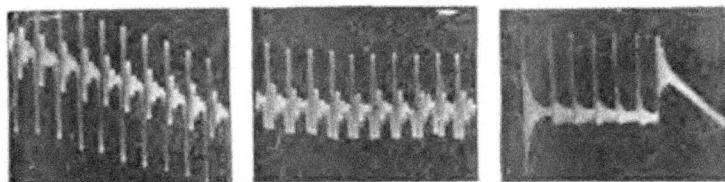

I.　　　　II.　　　　III.

FIG. 23.

---

[1] I have argued elsewhere what cannot be argued here, viz.,
that the positive effect above referred to has not been produced
by an ordinary Anelectrotonic current.

Let us now take nerves of these three stages, giving respectively the three types of electrical response just enumerated, and examine upon them the alterations of response brought about by carbonic acid.

The alterations produced in fresh nerve you have already witnessed. The negative variation was augmented. Those produced in transitional and stale nerve, especially the latter, are under all circumstances more uncertain ; we cannot unfailingly obtain a variation of the desired type, and even when we have obtained it—we cannot confidently predict the alteration that may result from our interference. And this is the worst month (February) in which to attempt to make such delicate demonstrations.

But the records of past experiments are of equal value with the experiments themselves, and the examples already alluded to in my second lecture, and to which I now return, are representative of the regular results of several hundred experiments, *all* the records of which are indexed and open to your inspection.

Here, then, is the effect of carbonic acid upon nerve of the second stage (fig. 26). The negative effect is increased, the positive after-effect is diminished.

Here is the effect of carbonic acid upon nerve of the third stage (fig. 28). The positive effect is abolished and replaced by a negative effect.

Here is another and less pronounced effect of carbonic acid upon nerve of the third stage (fig. 30). The positive effect is diminished.

Compare now with these three typical and characteristic effects of carbon dioxide, the effects produced upon nerve—either upon the same nerve, or at least upon nerve in a similar state—by prolonged artificial activity. You will recognise that in each separate case the effect of activity is similar to the effect of carbon dioxide. Considering the conditions of the case—that the *a priori* probable product of activity is carbon dioxide, we are, I think, entitled to conclude that the similar series of effects has been due to a common cause—to known CO_2 introduced from without in one series; to the hitherto unknown CO_2 evolved from within in the second series. The nerve itself has served as the reagent indicative of the presence of the latter.

Here is the nerve giving a small negative deflection followed by a large after-deflection, altered as you have seen by CO_2 from without, and similarly altered in consequence of tetanisation. (Compare figs. 26 & 27).

Here is the nerve giving a positive deflection, reversed to a negative deflection by CO_2 from without, and here is the same nerve exhibiting a similar reversal in consequence of tetanisation. (Compare figs. 28 & 29).

Here, finally, is the nerve of which the positive

First Stage

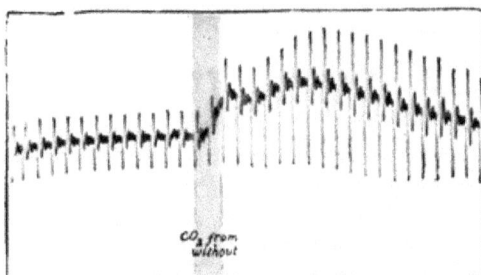

FIG. 24 (869).—Augmented negative effect.

Second Stage.

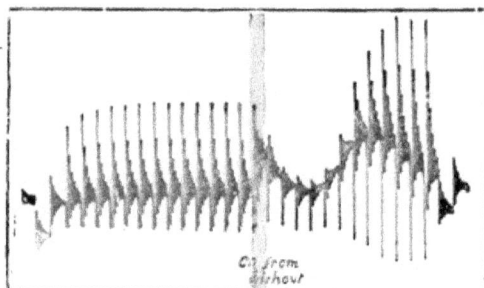

FIG. 26 (710).—Augmented negative effect. Diminished positive after-effect.

Third Stage.

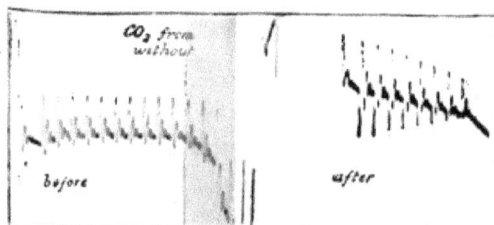

FIG. 28 (859).—Positive converted into negative effect.

Third Stage.

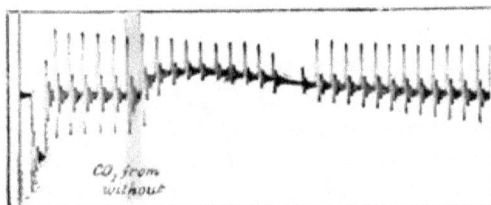

FIG. 30 (985).—Positive effect diminished.

INFLUENCE OF CARBON DIOXIDE ON THE ELECTRICAL RESPONSE OF NERVE
IN ITS THREE STAGES.

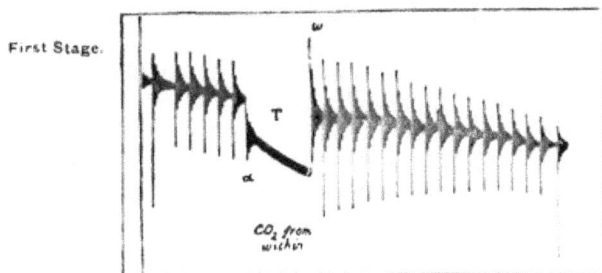

First Stage.

FIG. 25 (675).—Augmented negative effect.

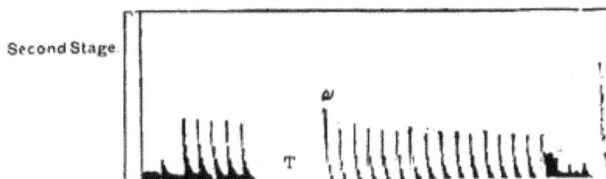

Second Stage.

ERRATUM.

p. 65. Description of Fig. 29 (858), for negative after-effect read negative effect.

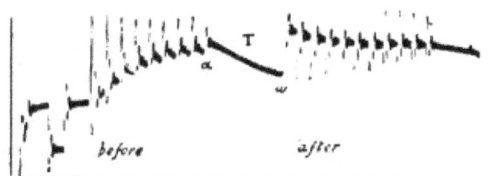

FIG. 29 (858).—Positive converted into negative after-effect.

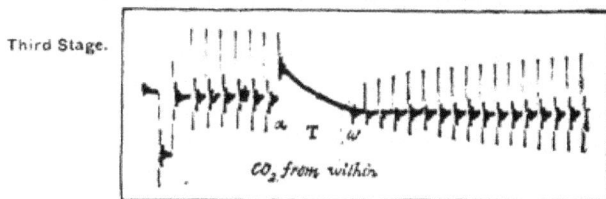

Third Stage.

FIG. 31 (984).—Positive effect diminished.
INFLUENCE OF TETANISATION ON THE ELECTRICAL RESPONSE OF NERVE IN ITS THREE STAGES.

5

deflection is only diminished by CO_2 from without, and here is the same nerve giving a similar diminution as the consequence of tetanisation. (Compare figs. 30 & 31). Indeed, if this last plate had been first brought to your notice, and without reference to the group of considerations just traversed, you would very likely have taken the diminution to be a common fatigue effect.

Provisionally admitting as an established datum that carbon dioxide is one of the products of nerve activity, our next question is : what becomes of it ?

I can venture upon no positive answer to this question. For the moment it seems to me that there are two possible answers open to our further investigation.

It is possible that the CO_2 may be dissipated by diffusion from the nerve, but it is also possible that it may be reintegrated within the nerve itself.

I have not yet found means of deciding between these two alternatives, which I have nevertheless not felt it unprofitable to state and briefly consider.

Obviously the first alternative presents itself first ; what more natural fate can we imagine for CO_2 if produced within a nerve, than its dissipation by diffusion into the surrounding atmosphere or into the lymph and blood, as in the case of the CO_2 evolved within a muscle ? I have nothing to say against this obvious probability, and can only point out as a remote possibility, suggested indeed by one or two peculiar features in the behaviour of isolated nerve, that CO_2 evolved in the process of disintegration may conceivably be reinvolved in a process of reintegration ; that, in short, baseless as the idea may seem at present, there may be in an animal tissue an assimilation of CO_2 analogous with the assimilation of CO_2 taking place in vegetable protoplasm. We have reason to associate the negative effect with a dissimilatory evolution of CO_2, we may some day find reason to associate the positive after-effect with an assimilatory reinvolution of CO_2. But at present this is a mere flying conjecture for which I have no positive base.

Putting this conjecture aside, let us turn to another less imaginary, yet still distinctly conjectural point.

Contractile tissue —that of the heart in particular, but also the contractile tissue of jelly-fish, as well as voluntary muscular tissue — exhibits a peculiarity known to physiologists as the "staircase phenomenon." Stimulated at regular intervals, not too long nor too short, by strong induction shocks, such contractile tissue gives a series of responses, each of which is the greatest effort of which the muscle is capable at the time, but each of which is a little greater than its predecessor. Such an increasing series is called a staircase, and we have seen that a series of electrica responses of nerve exhibits a similar staircase — increasing in the case of a series of negative effects, decreasing in the case of a series of positive effects or after-effects.　Remembering that the characteristic effect of CO_2 from without, and presumably from within, is an augmentation of negative response and a diminution of positive response, we have no difficulty in admitting that the electrical staircase of nerve, whether ascending or descending, is brought about by carbonic acid evolved at each step in a series.　It is pretty obviously a CO_2 phenomenon.

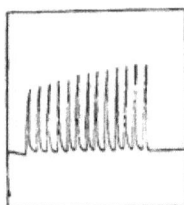

FIG. 32.—"Staircase" of contractions of a frog's heart.

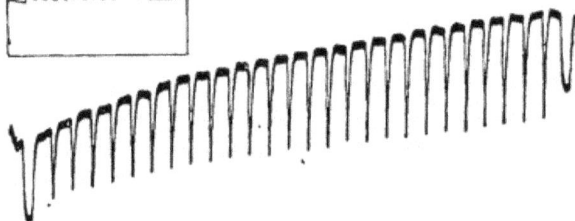

FIG. 32 A.—"Staircase" of electrical responses of nerve.

Staircase effect, then, is not confined to contractile tissue, it clearly applies to nerve where it is equally clearly produced by CO_2.　To my mind this precise notion is a welcome addition to our somewhat less definite psychological notions concerning staircase effects obtaining in central nervous action.

Summation of stimuli—*i.e.*, the gradual accumulation of a series of individually insufficient stimuli into an effective excitant—is itself conceivable as the expression of augmenting excitability by augmenting evolution of CO_2.　The establishment of paths of less

resistance in this or that direction of the central
nervous system—the "Bahnung" of German psycho-
logists — is conceivably due to a facilitation of
transmission along some line previously traversed by
nerve-impulses, and therefore with its excitability
sharpened by CO_2. We may even be tempted to
cast our imagination much further back—to the
very origin of specially conductile tissue within a
mass of homogeneous protoplasm. Picture to yourself a first linear discharge of action within such
mass ; carbonic acid is evolved along this line, and
the subsequent discharge of action will be in this
rather than in other lines. That will have been a
primitive form of "Bahnung" in the sense with
which Spencer's writings have made us so familiar,
and to my mind a knowledge of the relations between carbonic acid and nerve makes this idea more
concrete and tangible by suggesting a possible—I had
almost said probable—physico-chemical mechanism of
the result. And if there has been a disintegration,
there follows along the same line a reintegration of
matter, by which the nerve-path becomes organically
as well as functionally constituted. We know that
wear is followed, and more than made good, by repair ;
we also know that one of the products of wear is
carbonic acid. I wonder does this carbonic acid
become altogether dissipated : may it not perhaps be
reinvolved in some storage combination, as the nerve-
fat perhaps, that is so prominent a constituent of fully

evolved nerve. Such nerve consists of proteid axis
and fatty sheath ; the axis—which is the offshoot of a
nerve-cell—is the specially conductile part, the sheath
is a developmental appendix, not directly connected
with any nerve-cell. Yet, cut the nerve, and sheath
as well as axis undergo Wallerian degeneration,
which is evident proof of a functional commerce
between sheath and axis. You have seen further,
that such nerve is inexhaustible, yet that it exhibits
very clear symptoms of chemical change after action.
All these things, to my mind, reconcile themselves
with the notion that the active grey axis both lays
down and uses up its own fatty sheath, and that it is
inexhaustible, not because there is little or no expenditure, but because there is an ample re-supply.

This is wild hypothesis—an unbridled excess of
the imagination—and I shall be the first to admit
this, nor claim for it any value other than as a possible
motive for further trial.

Still, leaving aside the imaginary reinvolution of
CO_2 and the imaginary origin of specially conductile
strands, let me at least urge that the staircase effect as
a general phenomenon gains value in a definite and
concrete sense, both as a physiological and as a psychological idea, when we have admitted that carbonic acid
is a product of nerve activity and that carbonic acid
facilitates nerve activity. From a physiological standpoint it seems to me preferable to admit as an effective
factor of " summation " and of " staircase effect " and

of " Bahnung," that augmented mobility of protoplasm which is a characteristic effect of carbonic acid, the principal product of previous activity, than to say "that excitation arouses a tissue to a state of greater expectancy as well as of greater activity."

And in this connection one is tempted to at least ask oneself whether the converse phenomena known to us as "fatigue," the "refractory state," "inhibition," may not also be connected with the evolution of carbonic acid—whether its anæsthetic action, which is its full effect, may not come into play with exaggerated or with culminating activity. But upon the consideration of this negative aspect of nerve activity I do not feel able to enter at present.

In conclusion, let me briefly answer two questions that have been very frequently put me with regard to these records. I will answer them briefly, and by no means to my own satisfaction, for the answers are little more than counter-questions.

What interpretation do you place upon these negative and positive effects?

What is the meaning of that remarkable alteration of base-line caused by carbonic acid?

As to the first question, I ask myself whether the negative and positive effects are to be regarded as signs of opposite chemical movements, whether the negative is to be taken as a sign of disintegration and the positive as a sign of integration, in the sense of Hering's dissimilation and assimilation—or whether they are to be considered as algebraic sums of kathodic and anodic effects respectively—resultant from the predominance of one or other factor in the series of alternating currents used to stimulate the nerve. Further investigation will, perhaps, decide this point.

As regards the second question, I am very tired of it indeed, for no one fails to put it to me, and I do not know what the remarkable alteration of base-line means. Sometimes I get away from the question by saying that it means an alteration of the galvanometer zero, which is a pretty obvious "answer." If that does not answer, I have to apologise for the remarkable alteration, and to say that it has only been by reason of this uncompromising method of recording that it has been made so prominent, and that in ordinary galvanometric observations one does not attend to it. But that does not get me out of trouble, and I am told it ought to be attended to, that the rise or fall, as the case may be, are very significant. Yes, they are significant, but I don't know what they are significant of. I am quite aware of the fact that carbonic acid from without always drives the zero up at first, while carbonic acid from within (by tetanisation) always drives the zero[1] downwards, but I do not know what that means.

---

[1] By zero, I mean the position of rest of the permanently deflected magnet, not its position of rest when no current is passing.

Postscript.—The dubious tone of the remarks made on p. 62, arose from the fact that the rehearsal experiments carried out during the week preceding the lecture were so unsatisfactory that I abandoned all hope of obtaining any clear demonstration of the principal experiment. The nerves persistently said "check," giving only a very small augmentation of response in consequence of prolonged tetanisation, so small indeed as to be liable to escape detection on the demonstrating galvanometer, except to very close scrutiny. For this reason a recording galvanometer was put up in a dark room behind the lecture theatre, in circuit with the demonstrating galvanometer (as shown in fig. 10), so as to have a record of the experiment actually made, to be put into the witness-box at the end of the lecture.

As it turned out, however, the lecture experiment came out with remarkable distinctness; the response, after prolonged tetanisation (ten minutes), appeared to be about three times as great as before, and the photographic record was subsequently brought in as a somewhat superfluous piece of evidence.

This result, which was more surprising to the lecturer than to any of his audience, had been secured by Miss Sowton, who acted as lecture-assistant, and arose as follows :—

From previous experiments made in order to test the possible nutritive action upon nerve of proteids and of carbohydrates, we had found that lactose, among

others, appeared to have a considerable, if not absolutely certain, "nutritive" action. Miss Sowton had,
therefore, without my knowledge, made tetanisation
experiments the day before lecture upon nerves
allowed to soak in saline solution of lactose, and

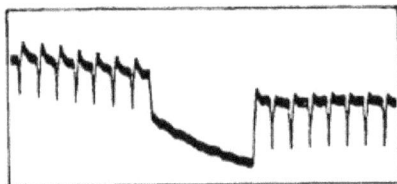

Fig. 33.   (2478.)   Effect of 5 minutes' tetanisation on an "unfed" nerve.

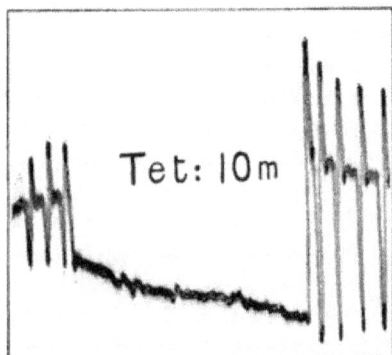

Fig. 34.   (2501.)   Effect of 10 minutes' tetanisation on a "fed" nerve.

had found that the typical effects in such nerves
were considerably augmented. The lecture experiment had been made upon a lactosed nerve, and,
like the experiments of the previous day, contrasts
very markedly with the other rehearsal experiments
made at this period on "unfed" nerves.

REFERENCES.

A full account of the subject of this lecture is given in *Phil. Trans. of the Royal Society* for 1897 (Observations on Isolated Nerve ; with Particular Reference to Carbon Dioxide. Croonian Lecture for 1896).

The " Staircase " phenomenon, first pointed out by Bowditch as being characteristic of heart-muscle, and by Romanes as occurring in the contractile tissue of Medusa, is considered by Romanes at some length from a general standpoint in *Jelly-Fish and Star-Fish*, International Scientific Series, p. 54.

The notion of " canalisation " or " Bahnung " is developed by Herbert Spencer in the *Principles of Psychology*, and alluded to by Romanes (loc. cit., p. 87), who gives references to earlier statements in a similar sense by Lamarck

LECTURE IV.

CONTENTS.

Polar effects.—In studying the electrical effects manifested by living matter, we shall repeatedly have occasion to employ electrical stimuli. It is important at the outset to avoid a perhaps natural confusion of ideas, and to expressly distinguish between electricity applied from without by way of what we shall designate as leading-in or exciting electrodes, and electricity arising within or aroused within the living animal or tissue, and conducted to the galvanometer or other electrical indicator by way of what we shall designate as leading-out electrodes. The latter alone is animal electricity, the former is not, although it is often used to arouse within living matter that chemico-physical action of which the deflection of a galvano-meter is an outward and visible sign.

This is, to some extent, a preliminary digression—inasmuch as the title of these lectures is Animal Electricity, not Electro-physiology. Strictly speaking, the former title covers only the electrical effects derived *from* animals, not the effects of electricity applied *to* animals. But seeing that we shall very shortly have to deal with polarisation phenomena which belong to both categories—being electrical responses to electrical currents; and that we shall have to examine the relation between such electrical responses and the ordinary mechanical responses significant of physiological excitation, the digression is not merely convenient and necessary, but logically defensible, even if we are to be restricted to animal electricity. The polar reactions of living matter are, as regards their physico-chemical mechanism, of the same nature as those of inert matter, but in many respects the former exhibit features that are peculiar, and characteristic of the living state. And while we must recognise that electrolytic disruption, whether of living, or of dead, or of inert matter, is of one nature in its essentials, we must also recognise that, in correspondence with varying conditions of greatly diminished chemical stability, varying degrees of greatly increased polarisability will be found in living as compared with dead matter. In the course of these lectures I shall show that the electrolytic polarisation of living matter is extraordinarily sensitive to chemical modifications that may certainly be termed slight, that, e.g., the electrolytic changes within a nerve are modified by irritant and by sedative drugs, and that a poison that kills protoplasm is, from our present standpoint, a reagent that immobilises the living electrolyte.

The questions arising in the consideration of the polar reactions of nerve are partly electro-physical, partly physiological, partly electro-physiological, and in this last respect distinctly overlap the narrower province of purely animal electricity.

Without, then, entering into details such as would be required under the heading of electro-physiology, and considering merely the general features of electrical excitation in so far as they involve electrical reactions and their relations to physiological reactions, these are the main points to be insisted upon.

It has been laid down by du Bois-Reymond that nerve is excited by a constant current, when that current begins and ends, i.e., at "make" and "break" but not while it flows, i.e., between the make and break effects.

It was thereupon further proved by Pflüger —

(1) That the "make" excitation arises at the kathode.

(2) That the "break" excitation arises at the anode.

(3) That during the passage of the constant current excitability is raised at and near the kathode.

(4) That during the passage of the constant current excitability is depressed at and near the anode.

These four propositions, specially applicable to nerve, express a very general law, and cover more particularly the case of muscle—of heart-muscle as well as of ordinary muscle. Subject to a possible exception in the case of non-fibrillated protoplasm, they express the law of response of living matter to electrical currents.

I have been at some loss how most briefly to present to you an experimental illustration of these two pairs of principles. The obvious and orthodox object of experiment, a nerve-muscle preparation of a frog, will not serve the whole purpose—for, while it would do well enough to exhibit augmented kathodic and diminished anodic excitability, it would afford no direct and evident proof of kathodic make excitation and of anodic break excitation.

I shall, therefore, have recourse to a less orthodox, and, in one particular, more complicated object, viz., a nerve-muscle preparation of a man, selecting for the purpose the ulnar nerve of my own forearm and the muscles to which it is distributed—these are, among others, several of the flexor muscles of the wrist and fingers.

Experiment.—I have connected myself with a battery by means of a large flat electrode at the back of the neck ; with the knob of the other electrode held by its insulating handle in my right hand, I feel about for the ulnar nerve at the back of the elbow; when the knob is felt to be comfortably applied, the current, directed so as to have its kathode at that spot, is gradually increased, the effect of its make and break being tested for occasionally. A strength is reached at which each make of the current gives a sharp flexor movement of wrist and fingers—*i.e.*, the kathode excites at make, or, otherwise, the make excitation is kathodic. Now the current is reversed, so that the kathode pressed down on the nerve is changed to an anode, and the current is made as before at the key, as you hear—but, as you see, without producing any effect; the anode does not excite at make. But neither does it excite at break—at least at this rather low current-strength. The current must be increased before any break effect appears with the anode over the nerve, and then there is also a make effect—which apparently contradicts the statement that the anode does not excite at make—that on the contrary it quells excitation.

The contradiction is apparent and not real, yet, as regards the course of our main channel of thought, any full consideration of the point would lead us astray. Let me then say rapidly that in this *embedded* nerve, current, having a fairly narrow way in (= anode), just under the electrode, has also a comparatively broad way out (= kathode) into the surrounding tissues, at some little distance from the actual electrode. The make effect is due to this extra-polar kathode—please do not regard it further; it is a complication due to the fact that the nerve is not isolated, but embedded.   Notice only that there is a contraction at break, which contraction is due to excitation of the nerve just under the anode, and let that signify for you that the anode excites at break, *i.e.*, that the break contraction is anodic.

These two statements—that the kathode excites at make and the anode at break—will be supplemented by the double statement that during the passage of a constant current, excitability is increased at the kathode, decreased at the anode—which is easily to be demonstrated.   And the clearest and least objectionable way to do this will be by an apparently very clumsy proceeding, viz., by light blows to the nerve, through the medium of the electrode itself, which is made anodic or kathodic at pleasure.   Pressing the electrode somewhat carefully upon the nerve, it is regularly tapped by a light mallet, just hard enough to give distinct twitches of the hand and fingers.   While the taps and twitches are proceeding regularly, a key is closed, rendering the electrode kathodic, and the twitches are evidently more pronounced, whereas with reversed current and an anodic electrode they are abolished.   This double experiment—of which the principal virtues are that it is simple, and that mechanical, *i.e.*, non-electrical, stimulation of perfectly normal nerve is employed—is good and sufficient evidence of the double statement that excitability is increased under kathodic influence and diminished

6

under anodic influence. Here is the record of an
experiment (fig. 35) that will serve as a memorandum of the fact. Anyone who may desire further
acquaintance with the doubts and complications of
which this very domestic-looking experiment is the

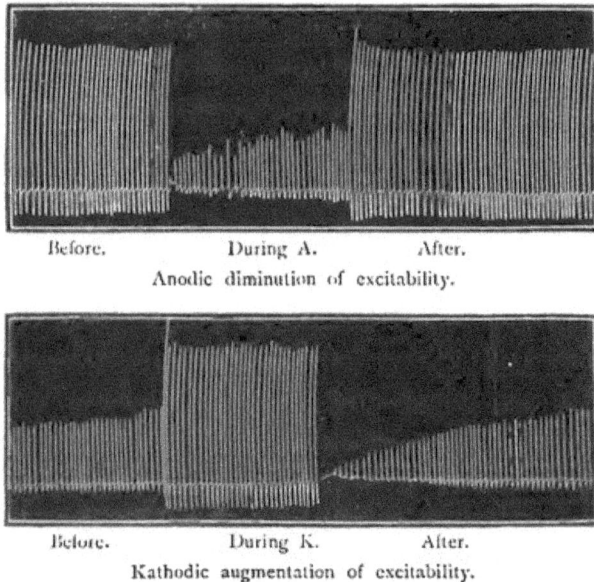

Before.   During A.   After.
Anodic diminution of excitability.

Before.   During K.   After.
Kathodic augmentation of excitability.

FIG. 35.—Influence of a polarising current upon the electrical excitability of
human nerve. (From Waller and de Watteville, *Phil. Trans. R. S.*, 1882).

settlement, may refer to the literature of the subject,
from which it may be gathered that a verification on human nerve of Pflüger's universally-admitted
principles *re* isolated frog's nerves was by no means a
matter of course, but had to be cleared before it was
clearly visible.

It is not my present purpose to enter into any detailed examination of this law of response ; that will form part of a future course of lectures. But it is necessary, in order to make clear to you the relation between the excitability and the electro-mobility of living matter, that I should allude to them now, if only to warn you that "excitability" and "electro-mobility" are not parallel attributes—that the term excitability is, in English, subject to an ambiguity that is avoided in the richer German by the terms Erregbarkeit and Leistungsfähigkeit, and that one of the several reasons leading me to adopt "zinc" as a new root word has been that its substantive "zinc-ability" (i.e., capability of being aroused to action analogous with that of the zinc of a voltaic couple) does not naturally lie wide open to ambiguity, as do the terms excitability and electromobility. By the terms "greater excitability," "more excitable," there may be implied "more easily excited," or "capable of greater reaction." By the terms "greater electromobility," "more electromobile" might be implied "more easily aroused to electromotive action," or capable of being aroused to greater electromotive action.

By the terms "greater zincability," "more zinc-able" it is most natural to understand—and at any rate it is the sense in which the expression will be used in these lectures—capability of being aroused to greater electromotive action, analogous with that of the zinc in a voltaic couple. That is the sense in

which the term "zincable" is to be understood in
the legend of this diagram (fig. 36), which is
intended to serve as a memorandum to the polar
alterations of excitability and of electromobility effected
by the action of the galvanic current.

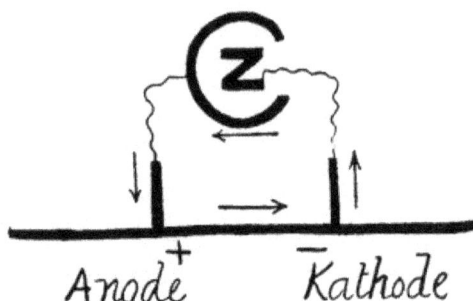

Fig. 36.
"*Excitability*" and "*Zincability*."

Less excitable.   More excitable.
Less zincative.   More zincative.
More zincable.   Less zincable.

We may tentatively proceed a step further in the
direction of general expression, and—admitting that
the grey axis is the essential part of a nerve-fibre—
recognise as a possible series of associated facts :—
(1) the kathodic excitation of the grey axis ; (2)
the liberation of electro-positive elements ; (3) the
reduction of deoxydisable "inogen" (? carbohydrate) ;
(4) the evolution of carbonic acid.

The point mentioned on p. 74. is in some measure
confirmatory of this view. The sugar appears in this
case to have played the part of an oxygen-giving food,

and it is conceivable that by the reducing action of living matter upon sugar, a deposit of fat might be effected.

There would at first thought seem to be very little continuity of ideas between the experiments we have just witnessed and those to which I am now turning. In point of fact there is a close relation between them ; the electrolytic disruption of simple chemical molecules, which we are about to touch upon, is in all probability the key phenomenon that will one day admit us to the deeper comprehension of the manifold disruptions and reunions of the complex chemical molecules that compose living matter.

It is a first step in that direction to make clear to ourselves that the phenomena of electrical excitation and inhibition are above all of polar and electrolytic (that is, of chemical) mechanism.   It will be a further step in the same direction to recognise that electrolytic movements, which stand out with comparative clearness from amid the unexplained jungle of excitatory phenomena—with such clearness indeed that we are at first sight tempted to deny their physiological character—are nevertheless subject to physiological conditions, and at the same time accessible to chemical modifications.

But I shall not apply myself to the building of a visionary castle—it is my purpose rather to diligently grovel among phenomena that are elementary and fundamental ; I allude more especially to electrotonic

currents, which underlie the electrotonic alterations of
excitability touched upon in the previous paragraph,
and which are undoubtedly the effects of electrolytic
polarisation.

Preliminary Experiment. — The two platinum
electrodes of a battery of two or three volts dip into
a rectangular glass vessel containing a mixed solution

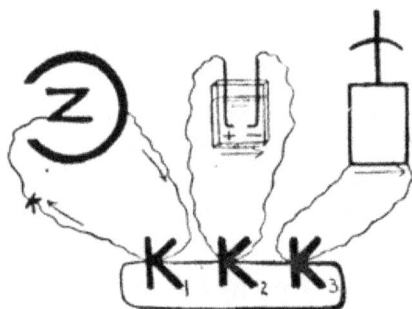

FIG. 37.

of dextrine and potassium iodide. The molecule of
the potassium iodide is composed of a basic moiety
potassium, and an acidic moiety iodine, and the latter
as soon as it is set free by the electrolytic disruption
of the molecule, will signify its presence by striking
a red colour with dextrine. [I have taken dextrine
in preference to starch because iodine strikes blue
with starch, and to most of us there is an association
of thought between redness and acid, blueness and
base.] The circuit is now completed, and at once
you see that the anode is becoming surrounded by a
red halo, indicative of the presence of free iodine, the

acidic moiety of the molecule K I. We may take this as our memorandum experiment signifying to us that with passage of current and consequent electrolysis there appear :

at the Anode	at the Kathode
acid	base
oxygen	hydrogen
chlorine	sodium
electro-negative ions	electro-positive ions.

These electrolytic products at the two poles respectively are themselves, so long as they are in *statu nascendi*, electromotive *against* the current that produces them ; thus oxygen, produced at the anode, and hydrogen, produced at the kathode, constitute a voltaic couple of which the current obstructs the original current by which they are generated, giving a secondary or polarisation counter-current that amounts virtually to a resistance against the original current. A voltaic cell in action, like living matter in action, surrounds itself with products of its own activity, obstructive of that activity. And the analogy that might be drawn up between the polarisation and depolarisation of a voltaic cell, and the disintegration and reintegration of living matter, would perhaps be not very far-fetched nor much more inaccurate than are all analogies.

But let me formally exhibit to you this fundamental fact of the polarisation counter-current. [It has indeed already exhibited itself informally in a way that

is rather instructive.   During the passage of the
original current the reddening of the fluid occurred
only at the anode ; but now—*i.e.* a few moments after
the original current has been cut out, and only the
counter-current has been in play—there is also a slight
reddening of the fluid round the other wire ; this
wire, which was kathode or way-out from liquid as
regards the original current, is now anode or way-in
to liquid as regards the counter-current.]

Experiment.—Beginning with K, closed (to cut out
the battery Ⓔ) and K, open, I first open and shut
the galvanometer key K, to see that there is no

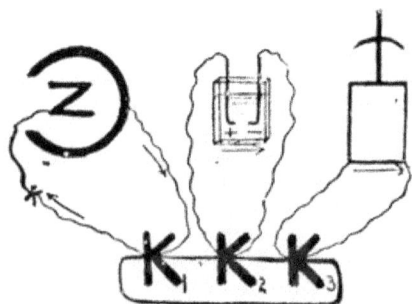

Fig. 37

current from the as yet unpolarised wires in the
polarisation cell. Leaving K, closed (to protect the
galvanometer from the comparatively large battery
current about to be used, and which will be
through the cell and would be through the galvanometer as shown by the arrows) I polarise the
electrodes for a moment by opening K,. Finally,
I switch off the battery by closing K,, and switch

in the galvanometer by opening K_2; the galvanometer indicates current against the direction of the
arrows, arising from the now polarised electrodes.
The key K_2 with which these are connected has
remained open throughout the experiment. The
electrodes have been so to say charged at K_1 from the
battery, and discharged at K_2 through the galvanometer, and, as you may have noticed, the magnitude
of the deflection has increased with the duration of
polarisation, and has diminished with the interval
between closure of K_1 and opening of K_2—*i.e.*, with
the time during which depolarisation has proceeded.

We seem to be a long way off from the case of a
living nerve, yet the distance between this case and
that of the polarisation cell is not so great as it seems.
Two more experiments will, I think, suffice to bridge
the gap—or at least to serve as stepping-stones
between the two orders of phenomena—from the
electrolysis that is purely physical to the electrolysis
that is at the same time physiological—from the disruption of inert matter to the disruption of living
matter :—

Of these two experiments one is intended to illustrate the fact—first established by du Bois-Reymond—
that the interface between different moist electrolytes
may be the seat of electrolytic polarisation ; the other
to illustrate the physical mechanism of the electrotonic
or extrapolar currents to which such polarisation gives
rise.

In illustration of the first of these two points I
cannot do better than show you an extremely simple
experiment that suggested itself only two or three
days ago as a possible lecture demonstration, and
of which nevertheless I already venture to say that
it does not fail. I only wonder how it is that I
never tried it before, for I have frequently wanted
an illustration of the kind, and wondered whether
the living tissues of the human body traversed by a

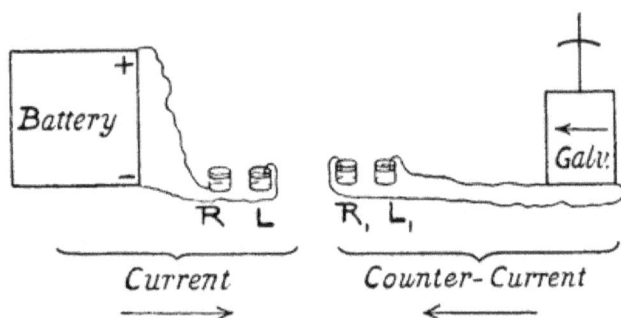

Fig. 38.—Plan of apparatus for the demonstration of "internal polarisation"
of the human subject.

galvanic current give rise to any demonstrable internal
polarisation, or whether they are so to speak kept
too well washed by the alkaline blood traversing their
capillaries.

Experiment.—Here are two vessels of salt solution
R and L connected with the two poles of a battery.
Here are two other vessels R' and L' of salt solution
connected with a galvanometer, which is shunted to
such an extent that on completing circuit through my

body by dipping a finger into each vessel, any acci-
dental inequality on the two sides does not affect the
instrument. Such being the case I polarise myself for
an instant by dipping a pair of fingers into the battery
vessels R L, and quickly transfer the same pair of
fingers to the galvanometer vessels, R' L' and as
you see, the spot is sharply deflected to the left. That
is a mixed effect of polarisation that may be wholly
external, at the way-in and way-out of current between
skin and fluid ; it merely serves to show the direction
in which to expect the effect of internal polarisation if
it exists. To this end—and this is the experiment
proper—I polarise myself again in the same way
through the same fingers dipped into R and L, but
for a little longer to make sure of the effect ; then I
put my possibly polarised tissues into the galvanometer circuit by dipping *another pair of fingers* into
the vessels R' L', thus avoiding the external polarisation of the first pair of fingers. There is now a
considerable deflection, not quite so sharp as before,
but in the same direction, independent of external
polarisation at the skin, and significant of internal
polarisation in the tissues below the skin. Finally
repeating the experiment with reversed direction of
current, a reversed effect is obtained. But to show
this without fail it is necessary to let the current flow
for a rather longer time in order to wipe out the effect
of the previous polarisation.

I shall not now enter into any details as to the

possible seat of this effect, which may obviously arise
at many sorts of interfaces between the various tissues
traversed by the current, but give the experiment as it
stands, in evidence of an internal polarisability of living
tissues taken *en bloc*. Obviously it gives no specific
information as regards the polarisability of any one
particular tissue.

The second of our two stepping-stones towards
the case of nerve is a purely physical experiment—
indirectly demonstrative of a peculiar distribution of

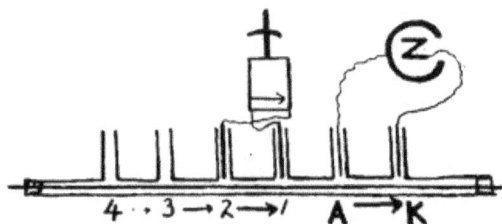

Fig. 39.—" Anelectrotonic " current from a core-model.

current, effected by polarisation, and characteristic of
nerve, which is just the one tissue among all other
tissues in which polarisation is most easily produced,
yet most difficult to directly demonstrate by reason of
its extreme evanescence. I know of no direct means
by which to demonstrate an internal polarisation of
living nerve. The chief evidence of polarisation is in
fact the peculiar extrapolar distribution of current
along a nerve, termed an electrotonic current, and it is
a similar extrapolar current that we are about to witness upon a polarisable core-model consisting of a
platinum wire surrounded by a fluid sheath of zinc
sulphate.

Experiment.—The original or polarising current [fig. 39] is from A to K. In consequence of electrolysis at the surfaces of entrance and exit between liquid sheath and metal core, counter-current is aroused, acting as a resistance, causing the surfaces of entrance and exit to spread along the model. In the figure you must imagine that the anodic or electropositive state shades off to the left from a maximum to a minimum value, so that if you connect a galvanometer with extrapolar points 1 and 2, then 2 and 3, then 3 and 4, you will

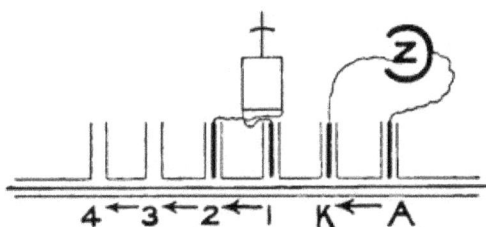

Fig. 40.—" Katelectrotonic " current from a core model.

get three diminishing values for current in the galvanometer from 1 to 2, 2 to 3, 3 to 4 (in the wire core from 2 to 1, 3 to 2, 4 to 3).

Precisely similar effects would be observed on connecting the galvanometer with a series of points on the side of the kathode to the right of the polarising current in fig. 40. But it will obviously be simpler to reverse the poles of the polarising battery, giving all currents with arrows pointing to the left instead of to the right.

These extrapolar currents, on the side of the anode in fig. 39, on the side of the kathode in fig. 40, precisely imitate what are known to physiologists as the Anelectrotonic and Katelectrotonic currents of nerve. And it is highly probable that the latter like the former are of electrolytic origin.[1]

But now as regards any possible future identification of ionic products in nerve, we must be on our guard. At the anode the current enters the fluid, then leaves it to enter the metal core ; the ions arising at the interface between sheath and core are thus kathodic or basic. Similarly the ions of the fluid sheath round the core under the battery kathode, are not kathodic, but anodic or acidic.

This point will come up again in the case of medullated nerve-fibres, where core as well as sheath is a moist non-metallic electrolyte. Meanwhile, towards the avoidance of a confusion not seldom made here between anode and kathode, way-in and way-out, I give a formal experiment on a core-model, in which

---

[1] Similar experiments can be made on core-models without any central wire, on e.g. a clay pipe soaked in salt solution and filled with a solution of copper sulphate ; du Bois-Reymond indeed showed long since that the surface of separation between two different electrolytes traversed by a current, gives birth to electrolysis (hydrogen and base in the electrolyte behind the current, oxygen and acid at the electrolyte in front). *Monatsber. d. Berl. Akad.*,17. Juli, 1856. Gesammelte Abhandlungen. Ueber Polarisation an der Grenze ungleichartiger Elektrolyte, vol. i., p. 1.

the anodic surfaces will be made visible to us by their acidic character. In the experiment illustrated by fig. 41, the battery electrodes (of platinum) are at 1 and 4, the electrodes between fluid sheath and platinum core are at 2 and 3. Taken in order we have anode to fluid at 1, kathode from fluid at 2, anode to fluid at 3, kathode from fluid at 4. The fluid is a mixture of dextrine and potassium iodide, soon after closure of the battery current the effect of anodic electrolysis is apparent as a reddening at the two surfaces 1 and 3, and you will not fail to remark

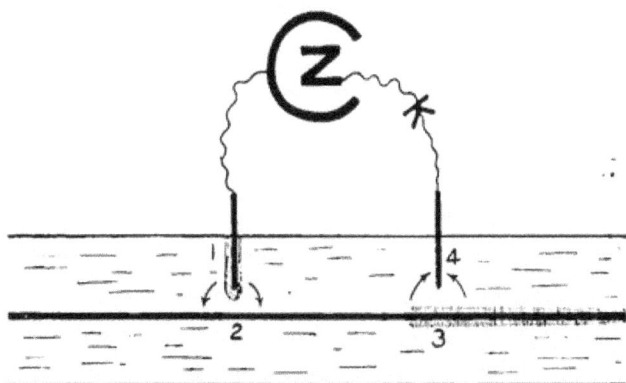

Fig. 41.—Experiment illustrating the entrance and exit of current to and from the sheath and axis of a core-model. The current in the direction 1, 2, 3, 4, enters the fluid at 1 and 3, leaves the fluid at 2 and 4. 1 and 3 are thus "anodic," 2 and 4 are "kathodic" as regards the fluid.

that at the latter of these the effect is most pronounced immediately under the battery kathode, but shades off in diminishing degree to a considerable distance along the wire in an extrapolar direction.

The core-model used in the experiment of p. 92,

consisted of a platinum wire in a solution of zinc
sulphate, and in that case there was no marked
difference of magnitude between the anodic and the
kathodic extrapolar currents.

Here is another core-model, composed of a zinc
wire in a solution of sodium chloride ; in this case
the effects are unequal.   On closure of the polarising

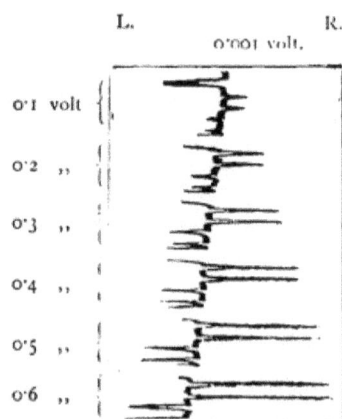

FIG. 42 (2369).—Extra-polar (= "electrotonic") currents of frog's nerve,
produced by polarising currents of increasing strength, from 0·1 to 0·6 volt.   At
each strength the A. and K. currents are taken twice.   Their magnitude may be
approximately estimated by reference to the standard deflection of 0·001 volt recorded at the commencement of the observation.

current in one direction (to the right), there is a well
marked anodic extrapolar current (to the right).   On
closure of the polarising current in the opposite
direction (to the left), there is a much smaller kathodic
extrapolar current to the left.

These two core-models reproduce to us what we
shall find to be the rule in the case of the nerves of

cold-blooded and warm-blooded animals respectively, and I will bring this group of introductory considerations to their conclusion by a cursory demonstration of this contrast—upon the nerves of a frog and of a kitten respectively—laying them in turn upon two pairs of electrodes by which polarising current is led into the nerve and extrapolar current is led out to the galvanometer.

With the frog's nerves the anodic extrapolar effect (to the right) is comparatively large, the kathodic

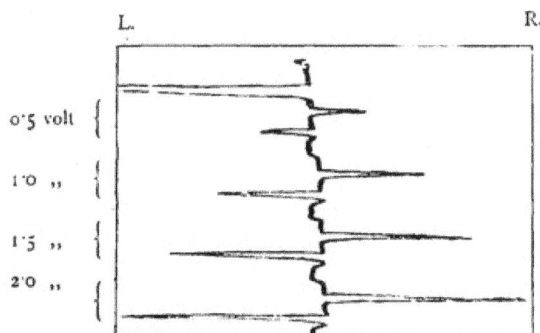

FIG. 43 (2383).—Extra-polar (== "electrotonic") currents of kitten's nerve, produced by polarising currents of increasing strength from 0·5 to 2·0 volts. (The standard deflection by 0·001 volt had a value of 40 mm., so that *e.g.* the A. and K. currents at 2 volts have an E.M.F. of about 0·0007 volt).

extrapolar effect (to the left) is comparatively small. These extrapolar effects are not due to current-escape, for they are, as you see, completely abolished by pinching the nerve between the two pairs of electrodes.

With the kitten's nerve the anodic and kathodic

7

extrapolar effects are well-marked and of equal
magnitude. They are completely abolished by
pinching the nerve between the two pairs of
electrodes.

These last experiments—that relating to frog's
nerve in particular—will form the point of departure
of my next lecture. We shall then become fully
convinced of the "physiological" nature of these
extrapolar currents—at least in the case of frog's
nerve. Of mammalian nerve I have little knowledge, and therefore, little to say. *Isolated* mammalian nerve gives no "negative variation," and I
was thereby deterred from taking it as an object of
systematic study. At present it is to me a rather
mysterious stranger.

Note.—Electrotonic currents, according to Biedermann,
are much less marked on non-medullated than on medullated
nerves, the katelectrotonic current in particular being absent.
This however is denied by Boruttau (Pflüger's Archiv, lxvi.,
p. 285, 1897).

The presence of electrotonic currents on non-medullated
nerves is in apparent contradiction with the view taken in
these lectures, that the interface between grey axis and fatty
sheath is the surface at which electrolytic polarisation takes
place. But the non-medullated state is not absolute, many
non-medullated nerves are more or less distinctly and
continuously myelinated, and in any case electrotonic
diffusion appears to be much less pronounced than on fully
myelinated nerves. There may prove to be some significance in the relation that apparently obtains in the com-

parative magnitudes of anelectrotonic and katelectrotonic effects in the different classes of nerves, viz., A. much greater than K. in non-medullated nerves, A. greater than K. in medullated nerves of cold-blooded animals, A. equal to K. in medullated nerves of warm-blooded animals.

But I am still in doubt concerning the " physiological " nature of the extrapolar A. and K. currents of isolated mammalian nerve, nor do I yet know whether the absence of " negative variation " on such nerves is in any way connected with the equality of these currents.

REFERENCES.

" Polar effects " were first fully described by Pflüger in his " Untersuchungen uber die Physiologie des Electrotonus." Berlin, 1859.

The polar effects on Man are described by Waller and de Watteville in the *Phil. Trans. R. S.* for 1882. (Influence of Galvanic Current on the Excitability of Motor Nerves of Man.)

Polarisable core-models were first systematically investigated by Hermann (*Pflüger's Archiv*, vols. v., vi., vii., 1872-3. *Handbuch*, vol. ii., p. 174). They have recently been still more closely studied by Boruttau. *Pflüger's Archiv*, 1894-6.

These three topics are briefly summarised in my " Introduction to Human Physiology," 3rd Ed., pp. 364, 366, 370.

Internal polarisation first alluded to by du Bois-Reymond in the " Thierische Elektricitat " in 1849, and again in 1883 in his monograph " *über secundär-elektromotorische Ers-cheinungen am Muskeln, Nerven und elektrischen Organen. Berliner Sitzungsberichte*, 1883.

In connection with the italicised words on p. 91, the following sentence is of particular interest : " Ich begreife aber heute nicht, warum ich nicht den Versuch so abänderte, dass beispielsweise mit den Zeigefingern der Schlag genommen, von den Mittelfingern die secundär-elektro-motorische Wirkung abgeleitet wurde." (Loc. cit., p. 371).

LECTURE V.

ELECTROTONUS.

" An." and " Kat."   Influence of ether and chloroform.
Physiological and physical effects. The electro-mobility
of living matter. Relation between polarising and extrapolar currents in nerve. Strength. Distance. Von
Fleischl's deflection.   Action currents are counter - currents. Extrapolar effects in mammalian nerve.

To-day's chief experiment presents to you a demonstration of du Bois-Reymond's electrotonic currents,
commonly referred to by physiologists as Anelectro-
tonus and Katelectrotonus, but which I shall often
take the liberty of calling by the shorter and more
familiar workshop names of An. and Kat., and by
the still shorter pen-names A. and K.

Experiment.   (Fig. 44.)   The nerve is resting
upon two pairs of unpolarisable electrodes, to receive
through *p p'* the *polarising current* from a battery (in
the present instance at the pressure of 1·5 volt), and
to give off through *e e'* the *electrotonic current* which
will be indicated by the galvanometer. The polarising circuit is completed at will by the key, and it is
made to pass in one or the other direction in the
nerve by means of a reverser, *rev.,*[1] parallel with the
transparent scale of the galvanometer.

---

[1] For prolonged experiments a revolving key is used, by
which the polarising current is made through the nerve at
regular intervals in opposite directions.

The nerve is facing you; and the connections are such that the direction in which the spot moves represents the direction of current through the nerve. Closing the key, I make a polarising current in the direction p p'; there is a considerable deflection to your right, indicating the presence of an extrapolar current in the nerve in the same direction from e to e', or *towards* the polarised region of

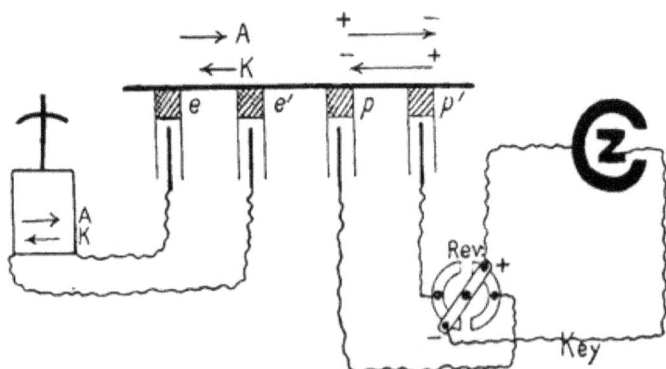

FIG. 44.—Diagram of experiment to demonstrate A. and K.

the nerve. This is an Anelectrotonic current, so-called because it is on the Anodic side of p p'.

Turning over the reverser to the left, and again closing the key, I make the polarising current in the opposite direction p' p; there is now a rather less pronounced deflection to your left, that indicates the presence of an extrapolar current in the nerve to your left, from c' to c, or from the polarised region p' p. This is a Katelectrotonic current, so called because it is on the Kathodic side of p' p.

You have now seen—and I will repeat the two trials in rapid succession — that in each case the extrapolar or electrotonic current, whether A. or K., passes in the nerve in the same direction as the polarising current by which it is aroused.

Your first thought, perhaps—as was mine when I first repeated the experiment—is one of disappointment. Is that all? Yes, that is the gist of the whole story, which you will find set forth at great length in a long series of memoirs extending over the last fifty years—at such length indeed that without some definite assurance of the simplicity of the facts, even a careful reader might pass it by, either missing the point altogether, or as was the case in every English text-book a few years ago, confusing it quite unnecessarily with the current of injury. Indeed, this confusion was made by du Bois himself at the very outset of his work, but the confusion that remained in his mind for at most a few months, was scrupulously preserved in the text-books for upwards of fifty years.

Your next thought is one of suspicion. Why this is mere current-diffusion along a conductor, just like what would happen along a hank of wet wool. That is a perfectly reasonable thought; ordinary current-diffusion might well take place; it is indeed a common fallacy ; we must in the first place observe all care to exclude it, and then we must take means of assuring ourselves that it has been excluded.

But even before this is done you will have noticed a point that hints at something in the nerve other than ordinary current-diffusion. The A. deflection and the K. deflection are unequal—A. is greater than K. If the two deflections had been by current-escape they would have been equal.

This does not indeed exclude all thought of current-escape, for the latter may be present with the peculiar something else that is becoming apparent to us. So we shall use further means to try the point.

Let us anæsthetise the nerve by a little ether or chloroform vapour, that will presumably distinguish between a " physiological " and a " purely physical " factor in the phenomenon, which *in toto* is, of course, physical. That which is " physiological " *i.e.*, dependent on the physico-chemical conditions peculiar to the living state will be suppressed ; that which is purely physical, *i.e.*, dependent on the physical properties of the dead nerve will persist.

Ether, Chloroform.—Here, then, are a couple of experiments in which this test has been applied, and from which you may recognise the propriety of distinguishing two factors in the entire phenomenon—a physiological factor, true An. and Kat., subject to anæsthetic influence—a purely physical factor, the

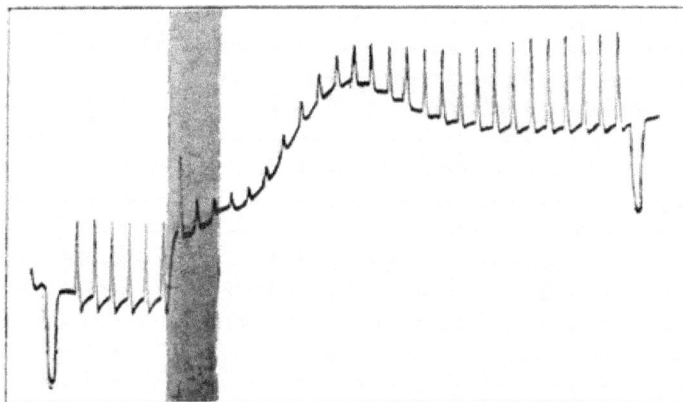

FIG. 45.—Effect of ether on the anelectrotonic current.

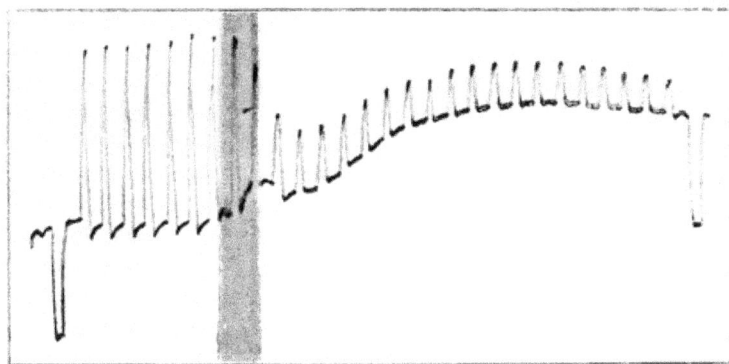

FIG. 46.—Effect of chloroform on the anelectrotonic current.

"physical¹ electrotonus" of German monographs, not suppressed by anæsthetics, which we shall designate by the colourless and non-committal expression "residual deflection."

This residual deflection—particularly well marked in fig. 46—itself gradually declines in course of time, mainly I think, by reason of the drying, and therefore increasing resistance, that you recollect to be one of the effects produced by anæsthetics. But this is a detail upon which we need not dwell, and in this connection it is hardly necessary to insist upon the obvious fact that in this case, as in that of the negative variation, chloroform exercises a permanent effect and ether a temporary effect.

There are other signs by which we can recognise that the extrapolar currents, An. and Kat., although

---

¹ How much sometimes turns upon a word! Hering, and his chief lieutenant, Biedermann, speak of "physiological" and "physical" electrotonus, and by physical electrotonus I for a long time supposed that they meant to designate a phenomenon intermediate between the physiological effect proper to living nerve and physical current-diffusion—an electrotonus in fact proper to dead, but otherwise anatomically perfect nerve. And in this belief, although explicitly stating that in my hands anæsthetics had served to distinguish between electrotonus and current-escape, I nearly succeeded in persuading myself of the existence of a physical electrotonus distinct from current-escape. Hering, however, has since informed me that "physical electrotonus" arose as a laboratory term meaning neither more nor less than current-escape.

in last resort physical, and, as we shall see, of electrolytic origin, are, nevertheless, entitled to the rather
indefinite qualification "physiological."

Interrupt the physiological continuity of the nerve
between the leading-in and leading-out electrodes, by
crushing it in the interpolar region c' p (fig. 44), or
better by touching it with a drop of strong acid. The
physiological conductivity is abolished, the physical
conductivity is intact (or enhanced if acid has been
used), but the extrapolar effects A. and K. are completely abolished.

Raise the temperature of the nerve above 40°, or
lower it to say—5°, and although—in the first case at
least—the physical conductivity has been enhanced,
the extrapolar currents, A. and K. are abolished, either
temporarily or permanently.

There is no time to-day for me to show you a
properly made temperature observation, as given in
fig. 63 (and it is only as I am speaking that it occurs
to me that we might have made a rough and expeditious trial by merely dropping the nerve into some
hot salt solution), but I can at once show proof that
the A. and K. effects you have just witnessed did not
depend upon current-escape. Pinching the nerve
with forceps, and then testing as before, we find
that they are completely abolished. There is no
deflection whatever, i.e., no current-escape.

These currents depend therefore upon the physiological state of the nerve —upon its "vitality"—and I

may add, vary with variations in that state.   Bad
nerves give bad currents, and that is specially the
case at this season of the year (February) when
nerves are at their worst.

There can be, I think, little doubt that these
extrapolar currents are an effect of electrolytic
polarisation.   This interpretation, which was origin-
ated by Matteucci, elaborated by Hermann, and
more recently by Boruttau, has completely displaced
the original interpretation of du Bois-Reymond, who
discovered the facts — so completely indeed that I think
its consideration could only complicate the question
to us ; if you are curious in the matter, you will find
it argued at length in the " Thierische Elektricität,"
vol. ii., p. 289 to 389.

Nerve, made up of medullated fibres, is an
electrolyte, or rather a pair of electrolytes.   Each
fibre consists essentially of a central grey core surrounded by a sheath of white matter.   And from
the fact that nerve composed of such fibres is the
only[1] tissue of the body that exhibits the extrapolar
currents just described, we may conclude that the
electrolytic interface is the surface of separation between grey axis and white sheath.

The electrolytic effects are as follows : current
entering the nerve by the anode, traverses the white
sheath to the grey core ; at the interface between

---

[1] Subject however to some reservation, as indicated in the note to Lecture IV.,
p. 98.

sheath and core it has its exit from sheath where it
liberates the positive ions (base, hydrogen), and its
entrance into core where it liberates the negative
ions (acid, oxygen, &c.). Leaving the nerve by the
kathode, the current passing from grey core to white
sheath has its exit from the core and its entrance
to the sheath, and consequently positive ions are

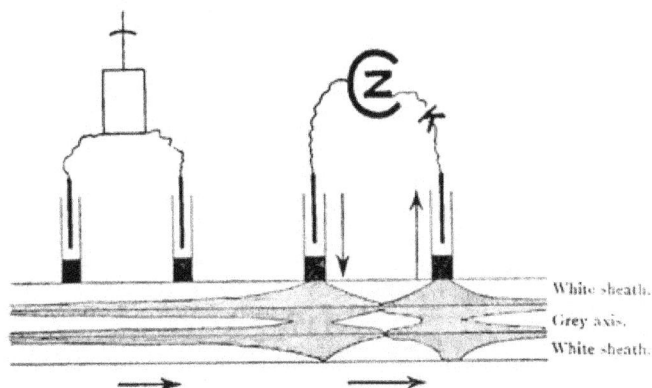

FIG. 47.—Diagram illustrating the theory that extra-polar currents of medullated nerve are due to electrolysis at the interface between sheath and axis.
Following the direction of the polarising current from copper to zinc through the
nerve fibre we have :

(1) Anode to sheath	Acidic electrolysis.		
(2) Kathode from sheath	Basic	,,	horizontal shading.
(3) Anode to axis	Acidic	,,	vertical ,,
(4) Kathode from axis	Basic	,,	horizontal ,,
(5) Anode to sheath.	Acidic	,,	vertical ,,
(6) Kathode to sheath	Basic	,,	

liberated in the former, negative ions in the latter.
The counter-current from these liberated ions
weakens the original current, is in fact tantamount

to added resistance at the interface, by reason of which a longitudinal spreading of current takes place on each side of each electrode. The extrapolar anodic and kathodic currents witnessed at the commencement of the lecture are thus accounted for.

Again you say perhaps "but this is physical, not physiological"; to which I should reply: "yes certainly it is physical, but that does not mean that it is not physiological." It is physiological in so far as it depends upon physiological conditions, upon the state of nerve that we call living, a state of peculiar physico-chemical lability, subject to all kinds of modifying influences temperature, moisture, drugs, all the influences in short that we are accustomed to consider in terms of their unanalysed effect on living matter.

Yes, the phenomena are physical; but they are also physiological: the two terms are not mutually exclusive, unless we reserve the term physiological for phenomena of which we are unable to detect the physico-chemical mechanism. The chief aim of physiological study is to express physiological phenomena in terms of physics and chemistry. And I venture to hope that we shall be able to push this enterprise a good deal further in this direction: I shall not be disappointed or surprised if much of our study of living nerve should turn out to be a study of the simple phenomena of electrolytic mobility, and if the depressant action upon nerve of anaesthetics and narcotics, and alkaloids, &c., should ultimately prove to be a chemical immobilisation of normally electro-mobile protoplasm. I do not, however, wish at this stage to pursue the argument further than may be necessary to convince you that it is the distinct aim of physiological endeavour to express the "physiological" in terms of the "physical" and "chemical."

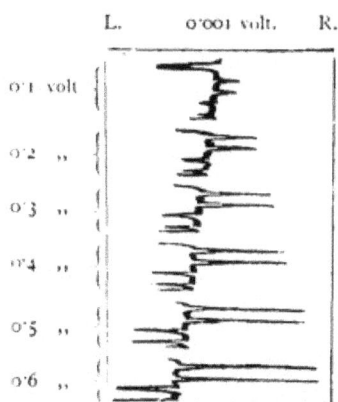

FIG. 48 (2369).—Extra-polar ("electrotonic") currents of frog's nerve, produced by polarising currents of increasing strength, from 0·1 to 0·6 volt. At each strength the A. and K. currents are taken twice. Their magnitude may be approximately estimated by reference to the standard deflection of 0·001 volt recorded at the commencement of the observation.

And it will be part of our immediate task to examine all the accessible electro-physiological phenomena of nerve with a view to determine whether or no they are reducible to the somewhat less mysterious form of electro-physical and chemical phenomena.

In this connection the next effects to be analysed
will be those first discovered by Bernstein and by
Hermann, viz., the electrotonic decrement and the
polarisation increment.

But before doing this I should like to take up two
or three further points relating to the electrotonic
currents themselves.

I shall recapitulate these points by means of a few
simple experiments, showing how the magnitude of
the extrapolar current varies ($1°$) with the magnitude
of the polarising current and ($2°$) with the distance
between leading-in and leading-out electrodes.

($1°$) The nerve is resting upon two pairs of electrodes p p' and e e' as shown in fig. 44. A polarising current of 0·1 volt gives rise to an electrotonic
deflection of 1·5 ; with the polarising current at 0·2
and 0·3 volt, the deflection is 3 and 4·5. From which
(and perhaps still better from the record reproduced
in fig. 48), we learn that the magnitude of the extrapolar current led off at e e' varies directly as the
magnitude of the polarising current led in at p p'.
This is quite in harmony with the view that the
extrapolar current is an electrolytic effect, and with
the fact that electrolysis varies directly with current
density.

Now for the effect of distance between leading-in
and leading-out electrodes.

($2°$) A nerve is resting upon six electrodes ; 1 and
2 are to remain throughout electrodes of the polar-

ising current, which is to be at the constant value
of about 1·5 volt; we shall lead off the extrapolar
current by 6 and 5, then by 5 and 4, then by 4 and 3,
that is to say, with distances between leading-in and
leading-out electrodes at 3, 2 and 1 centimetres.

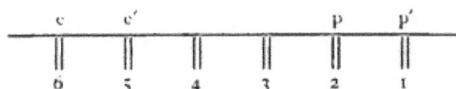

We begin with c c′ at 6 and 5, and complete
the polarising current through 1 and 2; the electrotonic deflection is just visible to you, being about
5 cm. of scale. We make a second trial with c c′ at
5 and 4; the electrotonic deflection is larger, about
30 cm. of scale. We make a third trial with e e′ at 4
and 3, the electrotonic deflection is much larger still,
right off scale, and to get a measurement of this deflection the galvanometer would have to be shunted.

From these data, and still more clearly from this
table, we see that the magnitude of an electrotonic

Distance between leading-in electrodes = 1 cm.
Distance between leading-out electrodes = 1 cm.
Polarising current by 1 leclanché cell (= 1·4 to 1·5 volt).

Length of nerve between leading-in and leading-out electrodes.	Anelectrotonic current.	Katelectrotonic current.	(*Negative variation by tetanisation.*)
3 c.m.	+ 1·5 mm.	— 0·5 mm.	(— 3·5 *mm.*)
2 ,,	+ 7·5 ,,	— 2·5 ,,	(— 4·0 *mm.*)
1 ,,	+ 120·0	— 90·0 ,,	(— 4·0 *mm.*)

current diminishes very rapidly with increasing distance between leading-in and leading-out electrodes.

8

This is, in fact, the one obvious distinction between an electrotonic current proper and a negative variation ; the latter is independent of the distance between leading-in and leading-out electrodes. No doubt both phenomena are an expression of polarisation, the former of a stable polarisation, the latter of an unstable and fugitive polarisation that propagates itself as a wave along the compound electrolyte[1] ; but at the present juncture it is not necessary that we should study this rather complicated question, and I shall reserve it for a future occasion.

I wish rather to exhibit some further definite data concerning the polarisation phenomena of nerve—data of which I cannot at present propose any final explanation, but which are evidently pieces of that particular puzzle.

One is a fact first pointed out by v. Fleischl, that has figured in physiological literature for several years past as a sterile item. The other is a little bunch of facts that have tantalised me for the last ten years by their paradoxical bearing. I have not hitherto ventured to call attention to them, and I do so now in a very dissatisfied frame of mind. They are fragments, evidently significant, but I do not know what they signify, I cannot even imagine what they signify.

---

[1] Boruttau has placed the finishing touch upon the identification between the phenomena in nerve and in core-models by showing that in the latter as in the former a diphasic negative wave is aroused by a single induction shock whatever the direction of the latter.

Here is von Fleischl's experiment.

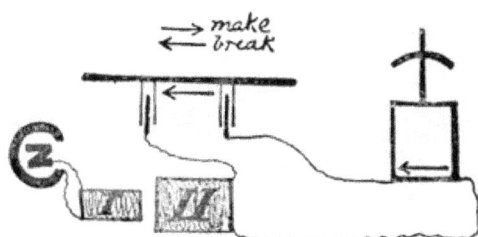

Fig. 49.

Nerve, galvanometer, and secondary coil are in one circuit. Without the nerve in circuit (*i.e.*, with a short circuiting key down) I test for the direction of the make and the break induction currents in the secondary coil and galvanometer. The make deflection is to your right, the break deflection is to your left, and, as you see, the two deflections are equal and opposite, so that if a series of rapidly alternating make and break currents is passed through the galvanometer there is no deflection. But when I open the nerve-key so as to put the nerve in circuit, keeping up the alternating currents, there is a deflection to your left, *i.e.*, in the direction of the break current.

To what is this deflection due? Is it physical or is it physiological? It is in any case due to a polarisation counter current; a few years ago I should have answered that it is physical, not physiological, and I might have illustrated the answer by reproducing the experiment on a polarisation cell without any nerve at all. But I should have had to take stronger currents,

and with various electrolytes I might have shown
deflection, now in the direction of the make, now in
the direction of the break. But now I answer that
it is physiological, and shall illustrate this statement
by killing the nerve (dipping it into hot water and
replacing it on the electrodes), and showing that on
the dead nerve there is no longer any deflection in
the direction of the break when the nerve is traversed
by the same series of alternating currents. As you
see, the galvanometer spot remains unmoved, even
when I considerably strengthen the induced currents ;
it is not till these are raised to an excessive strength
that the spot begins to shift. This is a physical
effect which I do not understand, and shall therefore make no attempt to explain. It is however
easy to distinguish from the deflection afforded by
living nerve.

Well then, admitted that the deflection in the
direction of the break is "physiological," present with
the living nerve but absent from the dead nerve, and
dependent upon the polarisability of living nerve —
are we able to proceed further in our explanatory
analysis ?

Not much further, with any legitimate assurance ;
v. Fleischl gives a highly complex explanation which
I shall not reproduce to you.[1] Hermann regards it

[1] v. Fleischl's experiment and his interpretation of it are
given in the " Wiener Sitzungsberichte, 1878.

as a polarisation increment.[1] I am not satisfied with
either of these explanations, but think it more probable
that the v. Fleischl's deflection which you have just
witnessed is a case of what du Bois-Reymond
designated as "positive polarisation," and Hering
and Hermann subsequently showed to be an after-
anodic action-current.[2] These are the several currents,
of which on this view only the after-anodic action-
current is apparent to you as a deflection in the
direction of the break. Seeing that the nerve and
galvanometer scale are facing you, and that the
connections are such as to make deflections on the
scale signify directions of current in the nerve (you

---

[1] Hermann, Handbuch, vol. ii., p. 167.

[2] There is still a fourth possibility, *i.e.*, that the effect in the
direction of the break may be due to a superiority of the
polarisation current after make over the polarisation current
after break. Although under the ordinary conditions of
physical experiment, the electrolysis by a break-current
exceeds that by the corresponding make-current, it is conceivable that the polarisability of living matter may be such
that the longer make-current may produce a greater electrolysis
than the shorter break-current.

The so-called "positive polarisation" or post-anodic action-
current, which is in the same direction as the exciting current,
is very readily intelligible as the effect of a post-anodic zincativity. It is easy to demonstrate. Its theoretical counter-
part during the passage of an exciting current, and opposed
to that current, is equally readily intelligible as the effect of
kathodic zincativity. It is, as far as I can see, insusceptible
of *direct* demonstration.

may indeed for the moment imagine that the scale
is an enormous nerve) we have :—

　1. The make-current to the right $+ \longrightarrow -$
　2. Its counter-current to the left $\longleftarrow$
　3. The break-current to the left $- \twoheadleftarrow +$
　4. Its counter-current to the right $\longrightarrow$
　5. v. Fleischl's current to the left $\longleftarrow$

i.e., as an after-anodic action-current of current No. 3,
which alone is exciting, and which gives rise to
"zincativity" on the side where the arrow-tail has
been figured. But for this arrow-tail representing a
post-anodic zincativity the four currents 1, 2, 3, 4
would neutralise each other, or if any difference
occurred it should be to the right by reason of an
excess of 4 over 2.

I am sorry to dwell so long upon this apparently
small and dubious point, still let me state in a few
sentences the reasoning that brought me up against
it. Considering that make excitation is kathodic
(p. 78), that excited tissue is zincative (p. 84), that
the make and break induced currents being equal
in quantity and unequal in potential, will therefore
be unequally excitant, while neutralising each other
on the galvanometer, I argued that with the nerve
and galvanometer in one circuit, it might be possible
to obtain a deflection in the direction of the make
owing to an action-current opposed to the break.
In the event I obtained just the opposite effect,
and reproduced what I recognised to be the effect

in the direction of the break previously observed by
v. Fleischl. My reasoning had evidently been im-
perfect : starting from the no doubt theoretically
correct assumption that any exciting current should
arouse an action-current of the nature of a polarisation counter-current, I sought to obtain evidence of
it upon nerve, and failed to do so, partly by reason
of the fact that the excited state in nerve does not
remain localised but spreads rapidly, partly by reason
of the then unknown after-anodic action-current,
which on the contrary appears to be a prolonged and
localised state. Direct evidence during the passage
of an exciting current of an opposite action current
did not and does not exist, yet from a theoretical
standpoint, it was and is a rather crucial point, and
I intend as soon as possible to try the point upon
more slowly reacting tissue. Meanwhile as regards
nerve, the nearest approach to direct evidence is
afforded by the " polarisation increment " to be
considered next ; this in my view is the negative
variation of a latent action-current opposed to the
polarising current. Indirect evidence is also afforded
by the abterminal and atterminal or abmortal and
admortal effects of Engelmann and of Hermann, as
will be developed in a future lecture.

Let me now ask your attention to another curious
and puzzling fact, concerning which I have no
glimmering of an explanation to offer.

Instead of a frog's nerve there is now a kitten's
nerve upon the two pairs of electrodes p p′ e e′ (fig.
44, p. 102); it is thicker than the frog's nerve, so
that our standard deflection by $\frac{1}{1000}$ volt is large, but
this is a detail. There is nothing remarkable about
the current of injury. I now test for the negative
variation in the usual way, and none is to be seen, although the resistance in the exciting as well as in
the galvanometer circuit is small, and although the

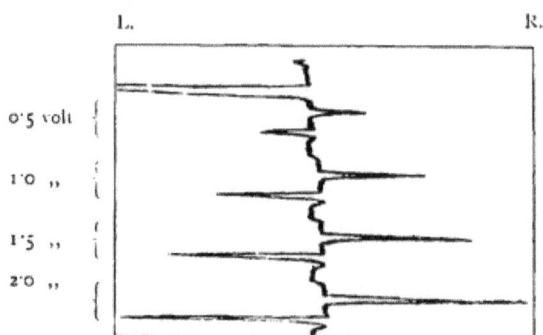

Fig. 50 (2383).—Extrapolar ("electrotonic ") currents of kitten's nerve,
produced by polarising currents of increasing strength from 0·5 to 2·0 volts. (The
standard deflection by 0·001 volt, not here reproduced, had a value of 40 mm.,
so that e.g. the A. and K. currents at 2 volts have an E.M.F. of about 0·0007 volt.)

strength of excitation be still further increased. This
has been no exceptional result ; I have never yet witnessed a true action-current from isolated mammalian nerve, placed upon the electrodes within a
few minutes after the death of the animal, whereas
frog's nerve under similar conditions has continued
to exhibit the action-current for hours and days
after excision. The contrast is glaring, and out of

all measure with the more enduring vitality of cold-
blooded as compared with that of warm-blooded
animals.

Turning our attention to the extrapolar or electrotonic effects, which are, as you have just seen, the
most obvious evidence of electrolytic polarisation
within the nerve, we shall speedily find ourselves
confronted by some paradoxical results. We obtain,
as you see, extrapolar currents that are equal and
opposite on the side of the Anode and of the Kathode
respectively, instead of greater on the side of the
Anode than on that of the Kathode as in the case of
frog's nerve. Of course you think of ordinary current
diffusion, but the extrapolar effects are not due to
ordinary current diffusion, for they are abolished by
crushing the nerve between the leading-in and leading-
out electrodes, or by dipping into hot water either the
end of the nerve that lies on the leading-in electrodes,
or the end that lies on the leading-out electrodes. So
we put the effects down in the category of physiological phenomena, and proceed to test them by
anæsthetic vapours, which as you remember produced
unmistakeable modifications of the extrapolar currents
of frog's nerves. The result is a little surprising and
by no means satisfactory ; the anæsthetics with which
we are most familiar—ether, chloroform, carbonic acid
—do not modify the extrapolar effects in the least.

And there the matter must stand for the present
as regards isolated mammalian nerve. No negative

variation.   Equal extrapolar currents, Anodic and
Kathodic, physiological as judged by the tests of
crushing and of hot water, not physiological as judged
by the test of anæsthetic vapours.   In short a
thoroughly unsatisfactory position of matters ; from
which, although I see no issue at this moment, an
issue must be found.   All that I feel entitled to say
at this perplexed stage is that you will at least realise
how it has happened that my experiments on nerves
that answered by "yes" and "no" have been many,
while on nerves that give no clear answer at all my
experiments have been few.   And perhaps at this
juncture the negative results afforded by mammalian
nerve may serve to throw into clearer relief the
positive results obtained on frog's nerve.

REFERENCES.

(1) *Electrotonic currents* discovered and named by du Bois-
Reymond in 1843 (" Thierische Elektricität," vol. ii., p.
289); explained as being polarisation effects by Hermann
(" Pflüger's Archiv," v., p. 264 ; vi., p. 312 ; vii., p. 301 ;
1872-3 ; Summary in " Hermann's Handbuch," vol. ii.,
p. 174).

(2) "*Positive polarisation*" first mentioned by du Bois-
Reymond in the " Thierische Elektricität," vol. i., p. 240.
and more fully described in the " Berliner Sitzungs-

berichte," 1883, p. 343 ; explained as being after-anodic action-currents by Hering in the "Wiener Sitzungs-berichte," 1883, p. 445, and by Hermann in "Pflüger's Archiv," vol. xxxiii., p. 103, 1884.

[These three memoirs are given in Burdon-Sanderson's "Translations of Foreign Biological Memoirs." Oxford, 1887].

Von Fleischl's deflection described in the "Wiener Sitzungs-berichte," 1878 ; considered by Hermann to be a polarisation increment ("Handbuch," vol. ii., p. 167).

(3) *Ether* as a means of distinguishing between "physical" and "physiological" electrotonus, described by Biedermann in the "Wiener Sitzungsberichte," 1888, p. 84 ; and in his "Elektrophysiologie, p. 693. Jena, 1895.

LECTURE VI.

ELECTROTONUS (*Continued*).

Influence of acids and alkalies. Influence of carbonic acid and
of tetanisation. Influence of variations of temperature.
The action-currents of polarised nerve. Bernstein's electro-tonic
decrement. Hermann's polarisation increment.

Influence of Acids and Alkalies.—Turning back to
the kind of nerve upon which we have found that
experimental comparisons can be made systematically,
let me next direct your attention to some experiments
that at once suggest themselves when we have admitted that the extrapolar currents of nerve are
caused by electrolytic polarisation.

Looking back to the series of effects represented
and summarised in fig. 47, p. 109, we naturally ask ourselves what will be the effects of acids and bases upon
An. and Kat.

Reviewing any considerable number of experimental records in which A. and K. have been taken
before and after the nerve has been submitted to the
action of a weak acid or of a weak base, we shall
find :—

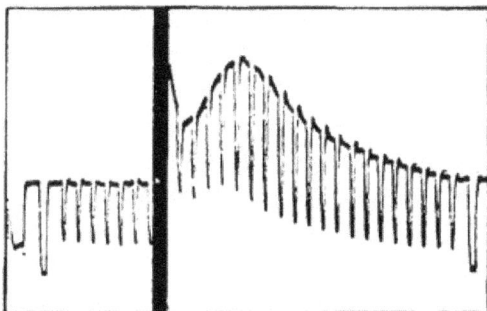

FIG. 51.—Effect of CO_2 on K, primary augmentation.

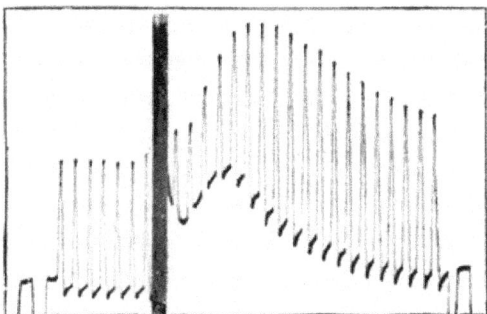

FIG. 52.—Effect of CO_2 on A, primary diminution, secondary augmentation.

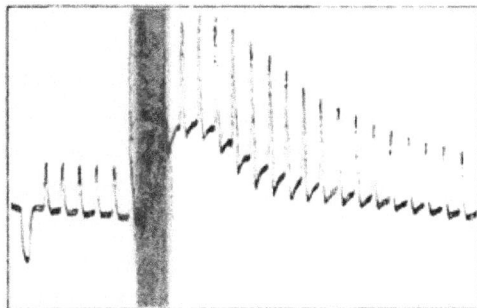

FIG. 53.—Effect of CO_2 on A, primary augmentation.

(1) That the most prominent and unmistakable modifications have been : an augmentation of Kat. by slight acidification ; a diminution of Kat. by slight basification.

(2) That An. has been sometimes augmented, usually diminished, but sometimes augmented by weak acidification, and

(3) That An. has been but little affected by weak basification.

These results do not sound particularly harmonious *inter se*, nor seem to be held together under any very obvious law. That depends, I think, upon the somewhat narrow range of concentration within which an acid or alkaline reagent effects what we may be entitled to designate as *characteristic* acid or alkaline changes.   Still we may even at this stage pick out as characteristic the augmentation and diminution of Kat. by acidification and basification respectively ; and we may remark that the augmentation of An. has always been effected by weaker acidification than the diminution of An.

A distinct step forward will be taken by the examination of records in which both An. and Kat. have been observed on the same nerve before and after its treatment with acid or with alkali. Within a certain range these reagents will have acted in opposite directions upon the two effects, or in the same direction but unequally, so that the ratio between their magnitudes is altered. We shall now find it convenient to adopt some conventional expression for that ratio. The number expressing the value of the magnitude of A. divided by the magnitude of K. presents itself as a natural formula, and we shall hereafter refer to it as the quotient A/K.

We are now able to say from examination of records in which both A. and K. have been modified, that the characteristic effect of acidification is a diminution of the quotient A/K, and the characteristic effect of basification an augmentation of the quotient A/K.

We are entitled further to make the statement that acidification affects first Anodic then Kathodic polarisation, causing first increased then decreased effects, their order of appearance being :

1. Increased A. as the first effect.

2. $\left\{ \begin{array}{l} \text{Diminished A.} \\ \text{Increased K.} \end{array} \right\}$ as the typical effect.

3. Diminished K. as the last effect.

The first effect is frequently missed, even with a weak acid, such as CO_2. The second effect is not often obtained pure ; more frequently we find a large diminution of A. together with a relatively smaller diminution of K., i.e. an absolute diminution of the quotient A/K.

These are dry statements ; it was, however, incumbent upon me to make them, in order to lay open the grounds upon which I have concluded that *the* *characteristic effect of acidification is a diminution of the quotient A/K, and the characteristic effect of basification an augmentation of the quotient A/K.*

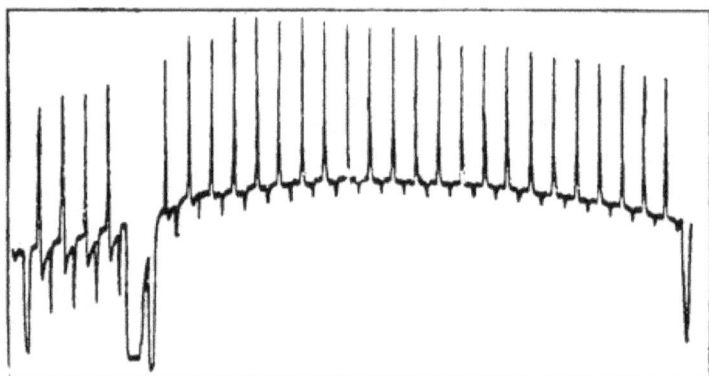

FIG. 54.—Effect of a weak alkaline bath (KOH, $N_{.0}$ or 0·285 per 100) upon A. and K.

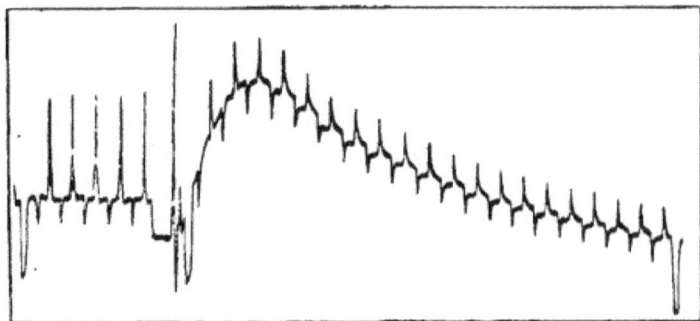

FIG. 55.—Effect of a weak acid bath (propionic acid $N/_{10}$) upon A. and K.

The influence of carbonic acid and of tetanisation.—

We have seen in a previous lecture that the nerve is altered by tetanisation precisely as it is altered by

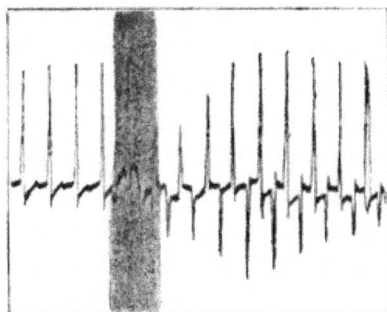

FIG. 56.—Action of CO_2 on A. and K.

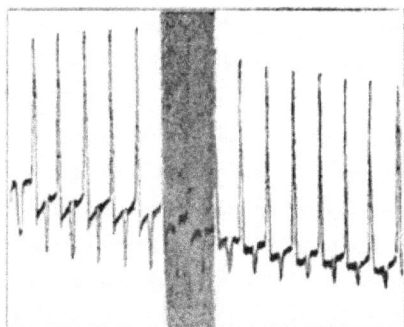

FIG. 57.—Action of ammonia vapour on A. and K.

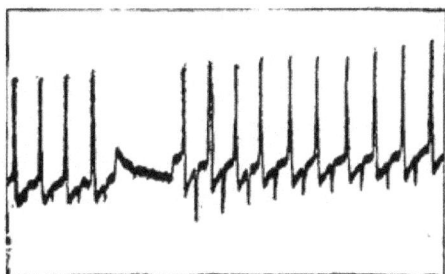

FIG. 58.—Effect of Tetanisation on A. and K.

carbonic acid, and we concluded therefrom that
tetanisation gives rise to an evolution of carbonic
acid within the nerve.

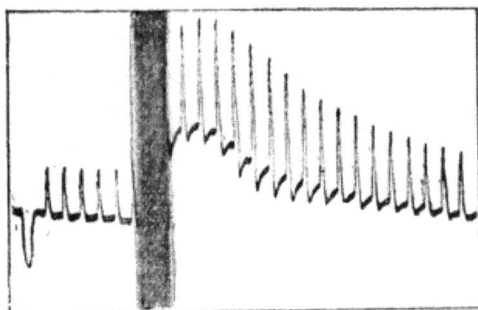

FIG. 59.—Effect of CO₂ on A. (primary augmentation).

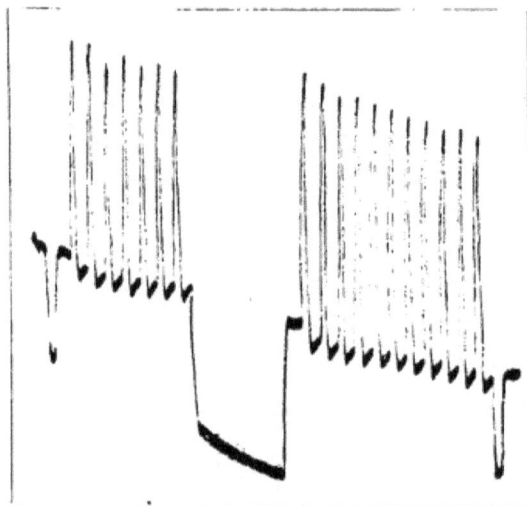

FIG. 60.—Effect of Tetanisation on A. (augmentation).

We are naturally led to test the effects of carbonic acid upon these extrapolar currents A. and K.,
which we have just found to be subject to modifica-

tion by ether and chloroform ; and as soon as we
have witnessed the effects of carbonic acid upon
these currents, our next obvious step is to test upon
them the effects of prolonged tetanisation.

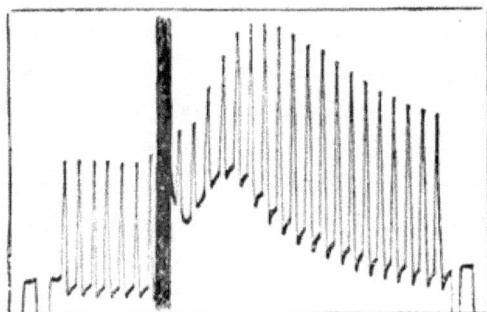

Fig. 61.—Effect of CO_2 on A. (primary diminution, secondary augmentation).

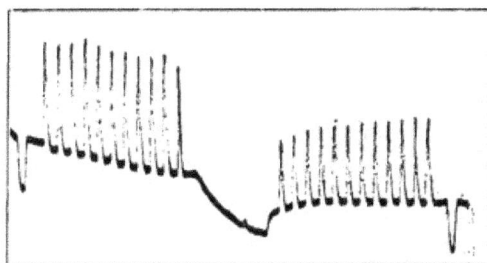

Fig. 62.— Effect of Tetanisation on A. (diminution).

The effects produced by carbonic acid—and I may
add, by tetanisation—are very characteristic. Their
description may best be given by the exhibition
of a series of representative experiments. Here, for
example, is a group of four such experiments, in

which the effects of carbonic acid and of tetanisation upon anodic currents have been recorded. Carbonic acid produces either an augmentation or a diminution of these currents. Tetanisation also produces either an augmentation or a diminution.

On reviewing a considerable number of records, we should find that in the case of carbonic acid, a diminution of A. is the rule, an augmentation of A. the somewhat rare exception ; whereas in the case of tetanisation, although a diminution of A. has usually been produced, an augmentation of A. has occurred frequently enough not to deserve the qualification of exceptional.

I think we may admit these results to rank as confirmatory evidence of the conclusion that tetanisation gives rise to carbonic acid ; an augmented A. is a slighter effect than a diminished A., and it is not surprising perhaps to find the slighter effect more frequent by tetanisation than by carbonic acid freely supplied.

But whether this be admitted as evidence or not —in Chapter III. the point is indeed already based on evidence which is, in my opinion, sufficient and conclusive — these experiments clearly show that the electro-mobility of nerve is modified by carbonic acid and by prolonged tetanisation.

The modifications of the Kathodic current effected by carbonic acid are more regular than are those of the anodic current. The former appears to be less

sensitive than the latter, and in all the experiments
I have made up to this time, an augmentation of K.
has been produced by carbonic acid and by tetanisation. This is illustrated in figs. 57 and 58, where
both A. and K. alternately produced have been
placed under observation. In both cases the K. current has been obviously increased by carbonic acid
and by tetanisation respectively, but the A. current
diminished by carbonic acid in fig. 57, exhibits no
marked alteration in consequence of tetanisation in
fig. 58. Presumably in this last case a mean happens to have been struck between augmentation and
diminution.

The Influence of Temperature. — It is a very
simple matter to examine the effect of a raised or
lowered temperature upon the extrapolar currents,
and the results—in the case of raised temperature
at least—are very constant and very characteristic.
All we have to do is to place the nerve-chamber in
a receptacle that is packed round with ice and salt,
or gradually warmed by a spirit-lamp, and then take a
continuous observation in the usual way. I shall not
tax your patience by asking you to witness such an
observation ; to be of any value it would have to be
· slowly made with the temperature raised or lowered
through a range of 20 degrees at a rate of a degree
a minute. Moreover, the photographic record gives
a better general view of matters than the observation
itself, and although less interesting perhaps, is cer-

tainly more trustworthy. Here is such a record
(fig. 63), of a series of alternating A.'s and K.'s at
one minute intervals, with the temperature gradually
raised from 18° to 40° and then allowed to fall. There
are two principal points exhibited in this record,
which is a typical one ; firstly the fact that at about
40° the extrapolar currents are somewhat suddenly
diminished almost and sometimes quite to extinction—

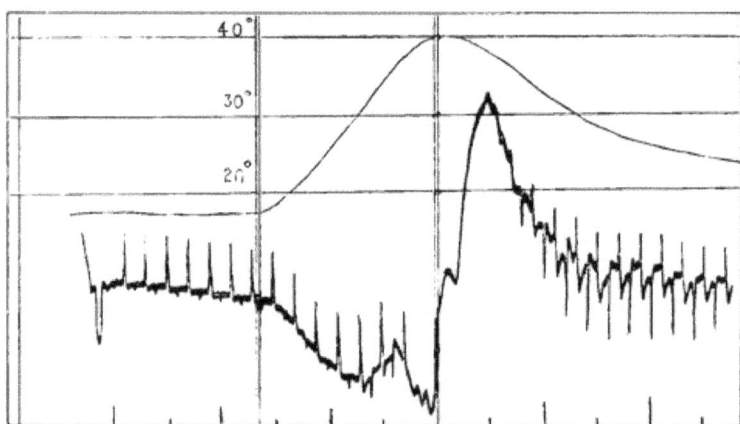

FIG. 63.—Influence of rise of temperature upon A. and K.

which is evidence of their physiological character ;
secondly, the fact that in consequence of the rise of
temperature A. has diminished and K. has increased.
This last is, I think, a very characteristic and note-
worthy point. For referring back to the last two sets
of experiments, relating to the effects of acidification
and of tetanisation, we find that all three of these

agents—acid, heat, and action—which presumably provoke a chemical disruption of living matter— have as their common effect an augmented kathodic polarisability.

And please notice with reference to this record, that there can be no question here of any fallacy by alteration of resistance. This point has of course been carefully examined by other experiments into which it is not necessary to enter now ; but in the present case, with a *raised* temperature (40°), *i.e.,* with *increased* conductivity, there are *diminished* currents, and the *increased* K. is with a *falling* temperature, and cotemporary with a diminished A. Comparing the relation between A. and K. before and after the rise of temperature, there is a very evident diminution of the A/K quotient ; as was the case in consequence of acidification, and in consequence of tetanisation.

The Electrotonic decrement. The Polarisation increment.—There are two more cardinal experiments belonging to our subject that I am anxious to demonstrate and to explain as clearly as I am able, and that will bring to its conclusion this preliminary exposition of the principal data concerning the electromotive reactions of nerve.

The *first experiment* is to show that an electrotonic current, like a current of injury, is diminished during excitation—undergoes a negative variation.

(Bernstein.) We shall see that this is true for both
directions of current—that both An. and Kat. are
diminished during excitation.

Let us begin with An., the connections being as
in fig. 64, with exciting and polarising circuits con-
joined as shown, so that both these currents pass into
the nerve by the same pair of electrodes.

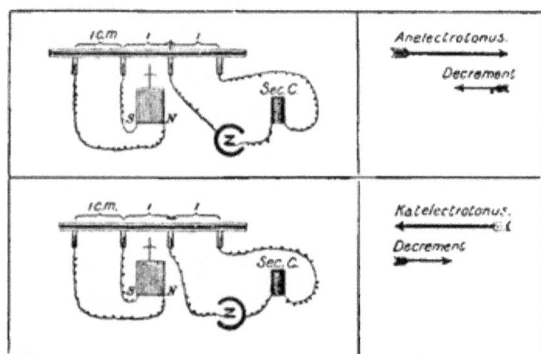

Fig. 64.

I first make the polarising current, which provokes
an anodic extrapolar current to the right in the nerve
and in the galvanometer; now that the deflection is
steady I excite the nerve, and during the excitation
the spot moves to the left, indicating a diminution of
the extrapolar anodic current.

How shall we best interpret these facts. Let us
first formulate them in terms of the usual signs + and
—, positive and negative. During polarisation p' is

p is +, e' is less +, e is least +.   In relation to each other e' is + and e is —.   Write these signs down then against e' and e of the figure (fig. 65) and trace the current through.   You probably first give it as being from e' to e, *i.e.*, the reverse of what it is, until a little reflection makes clear to you that as regards extrapolar current, e' is kathode and e is anode.   So you change the signs to + at e and — at e'.   If from this point you follow the change produced during excita-

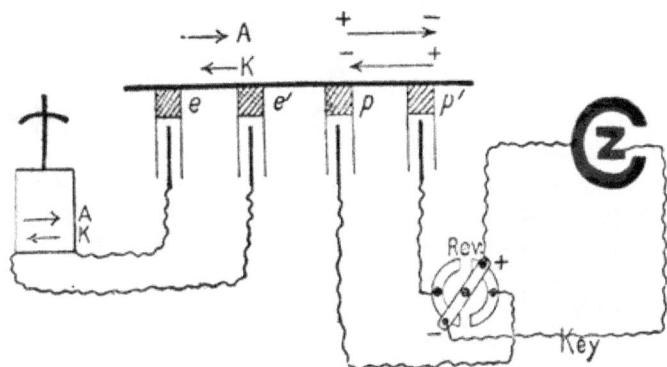

FIG. 65.

tion, you would not go wrong whether you take + and — potentials at 3 and 4 respectively, or their — and + poles at the galvanometer.   Still mistakes are often made, and it is necessary to be carefully on one's guard, for it to come out evidently that excitation must give an effect negative to the electrotonic current.

The way in which I prefer to express it is this : e' is more anodic than e, *i.e.*, less zincative, and current in the nerve is from e to e'. e' is more anodic than e, *i.e.*, more zincable, and the excitatory change gives current in the nerve from e to e'.

Our *second experiment* is to show that a polarising current is increased during excitation—undergoes a positive variation. (Hermann).

The connections are as in fig. 66, as being perhaps a little more obvious than would have been the case with polarising current, galvanometer and secondary

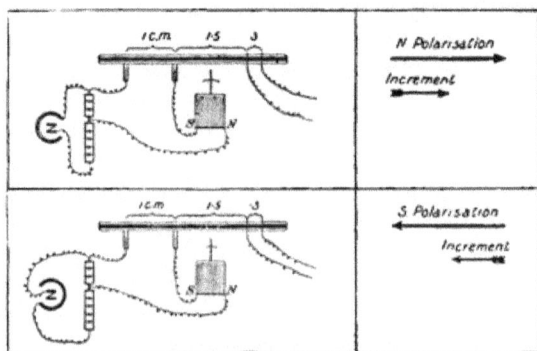

Fig. 66.

coil in one circuit connected with the nerve by a single pair of electrodes. [The resistance boxes r. R., used as shewn in figs. 10 and 11 on pages 33 and 38, enable us to take a convenient voltage (0·1 to 0·5) for the polarising current.]

Using a Leclanché cell and having set the resistances at 1000 and 14000 ohms respectively to get a voltage of about 0·1, I close the key in the polarising circuit giving current from e to e' in the nerve.

The spot flies off scale to the right.   I bring it back
on scale by a considerable movement of the controlling
magnet, and when it is steady, excite the nerve by
closing the key of the induction coil.   The deflection
during excitation is to the right, *i.e.*, the polarising
current is increased.

This is precisely as you might have foretold, never
having seen the experiment before.   The nerve
at c is anodic, zincable, and during excitation it gives
current from c to c', *i.e.*, with the polarising current.
You may perhaps express this in terms of positive and
negative, but by no means so clearly ; in fact this
simple matter under the title of "polarisation increment" is one of the well known posers of honours
examinations in physiology.   But I should not advise
any candidate to say "zincable" to his examiner, I
only advise him to think "zincable" in order to understand the subject.

Let us fix the matter by repeating the two experiments with a reversed polarising current.

In the first experiment (of the electrotonic decrement) p is Kathodic ; the electrotonic deflection
(Kat.) is from right to left; during polarisation the
nerve is less Kathodic at c than at c', *i.e.*, less zincative
and more zincable at c, and during excitation there is
current in the nerve from c to c', *i.e.*, the electrotonic
current is diminished.

The second experiment (of the polarisation increment) is obviously the same whichever way you take it.

Here is the experiment of the polarisation increment with a slightly different circuit. The polarising
current, galvanometer, secondary coil and nerve are
in one circuit. The nerve receives the polarising
and exciting currents, and gives its electrical response
through a single pair of unpolarisable electrodes.

The first thing to do is to see that there is little
or no current from the electrodes and portion of nerve
between—if the nerve happened to be injured near
one or other of the electrodes, we should have a
current of injury and on subsequent excitation a
negative variation of that current that might confuse
or mislead us.

The next step is to set the alternator in movement
and see that the induction shocks rapidly pulsating
to and fro through the nerve and galvanometer, do
not of themselves give any deflection. If the induction currents were taken too strong, we should be
confused by an initial and terminal kick by the first
make and last break shock of the exciting series, and
by a permanent deflection during the series in the
direction of the break current, to which allusion was
made above (p. 114).

These two preliminary sources of error having
been excluded, we may proceed with the experiment
proper.

I close a key in the polarising circuit. The spot
flies off to the right. I bring it back to scale by
means of the controlling magnet. And now that it

is at rest, I excite the nerve. There is a deflection
to the right, *i.e.*, an increment of the polarising
current.

Repeating the experiment with the reversed
direction of polarising current, everything is reversed, viz., this current is to the left, and its
excitatory variation is also to the left.

Trace out the rationale of the polarisation increment in this form, and you will recognise that in both
cases the anodic portion of nerve, or "way in" of
current, being the more zincable spot, becomes more
zincative during the superadded state of excitation.
There is thus "current of action" in the same direction as polarising current.

These varieties of effect in accordance with various
combinations of circuits, might readily be extended.
We might, *e.g.*, test for the decrements of An. and
Kat. with the coil in the electrotonic and galvanometer circuit, instead of in the polarising circuit.
Or going back to our second fundamental experiment, as described in the first lecture—the negative
variation of nerve current (p. 10)—we might combine
into one the leading-in and leading-out circuits.

The results of such experiments would come out
precisely as might be anticipated. Wherever current
enters a nerve from an electrode, the nerve is anodic
and zincable, whether such current be a polarising
current sent *into* the nerve, or an electrotonic current,
or a current of injury drawn off *from* the nerve.

But I have thought that the possible confusion of ideas resulting from such experimental complication might overbalance the advantage of the extension of data, and even the distinct practical advantage in certain cases, of being able to test short bits of living tissue through a single pair of electrodes.

Besides, if the principle has been mastered, its manifold applications can present no further difficulty; an allusion to them will have been sufficient, their detailed description would be superfluous and therefore tedious.

An action-current, however excited, and under whatever circumstances, whether as a negative variation of an injury current, or as a positive variation of a polarising current, or as a negative variation of an electrotonic current, is due to a physiological (= physico-chemical) inequality between two points. Active tissue is "zincative," resting tissue is "zincable."

Look at this last diagram. Is it not forbidding? It summarises all the currents experimentally detected in nerve; I shall not undertake to wade through their redescription in cold blood. To anyone who has not mastered their key, they must remain unintelligible; to anyone who has mastered their key, they will be a legible and symmetrical page in the story of living matter.

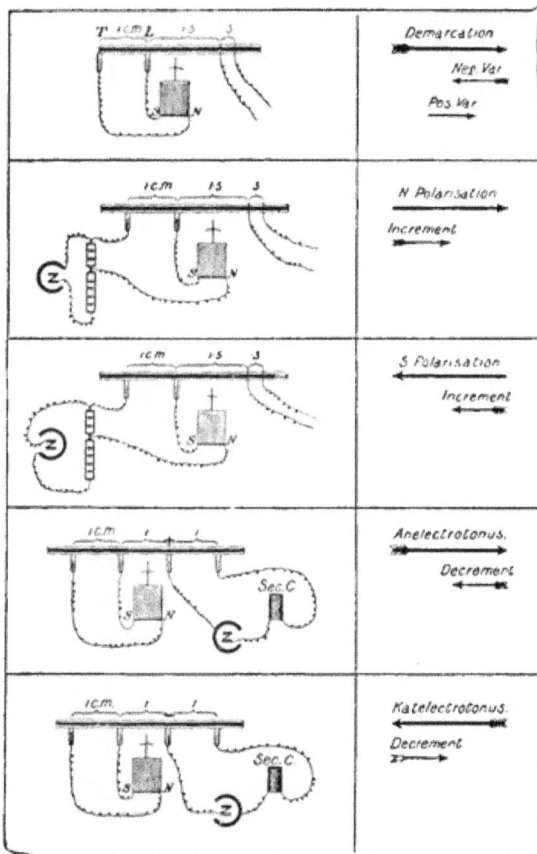

T 1 *c.m* *L* 1.5 *S*	Demarcation Neg. Var. Pos. Var
1 *c.m* 1.5 *S*	N Polarisation Increment
1 *c.m* 1.5 *S*	S Polarisation Increment
1 *c.m* 1 1	Anelectrotonus. Decrement
1 *c.m* 1 1	Katelectrotonus. Decrement

Fig. 67.

REFERENCES.

Electrotonic decrement first described by Bernstein in du Bois-Reymond's Archiv., 1866, p. 614.

Polarisation increment first described by Hermann in Pflüger's Archiv., vol. vi., p. 560, 1872; and further explained in Pflüger's Archiv., vol. vii., p. 349, 1873.

[Previously noticed by Grünhagen, but by him attributed to a diminution of resistance. (Zeitschrift für rationelle Medicin, 1869.)]

Experiments illustrating the action of anæsthetic and other reagents on electrotonic decrements and polarisation increments are given in the Croonian Lecture for 1896. (Phil. Trans. R.S., 1897).

Fig. 10.

General plan of apparatus employed to investigate the influence of
reagents upon the electrical response of isolated nerve.

The nerve-chamber contains the nerve resting upon a pair of unpolarisable
electrodes connected with K_1, and a pair of platinum electrodes, through which
the nerve is excited. The wash-bottle serves to prevent drying of the nerve, and
in certain cases to apply vapour.

The exciting apparatus is represented above; the circular interrupter in the
primary circuit revolves once a minute, and makes contact at a mercury pool for
that each revolution. The vibrating interrupter of the coil starts as the circuit is
completed, and the nerve is thus tetanised each minute for 25 seconds, at a fre-
quency of 70-75 interruptions per second.

The keyboard is composed of four bridging keys. K_1 in connection with the
nerve at I and I, k_2 with the compensator, K_3 with the recording galvanometer,
and K_4 on occasion with a demonstrating galvanometer. When all the keys are
closed everything is short-circuited through them; the keyboard is then simply
a metallic ring. Opening K_1 and K_4 puts the nerve into connection with the
galvanometer; opening k_2 lets into the nerve circuit any suitable fraction of a
volt from the cell and rheostats at R. To take tones volt from a Leclanché
cell, r is taken 10 ohms, and R 14,000 ohms; to take a petromillion current
of 1/10 volt, r is taken 100 ohms, and R 14,000 ohms, &c.

N.B. The galvanometer records its movements to Right and Left of
some arbitrary position of rest. The connections and disposition of the
apparatus are such that deflections to the Right or Northwards represent
positive effects, and deflections to the Left or Southwards represent
negative effects.

The completed records (with the exception of Figs. 42 and 43) are
turned so that positive deflections to the Right or North read upwards,
and negative deflections to the Left or South read downwards.

Thus − : the negative variation reads downwards, anodic deflections
read upwards, kathodic effects read downwards.

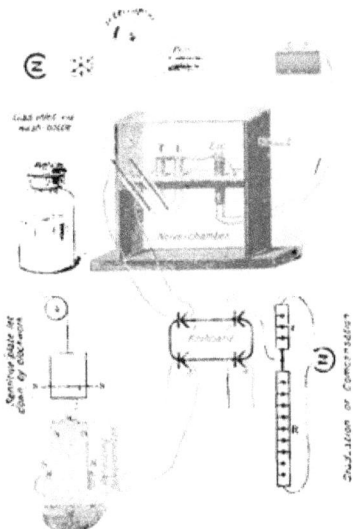